ACPL IT

DISCARDED

620.00
Wilhelm, Robert G.
Computer methods for
tolerance design

**DO NOT REMOVE
CARDS FROM POCKET**

ALLEN COUNTY PUBLIC LIBRARY
FORT WAYNE, INDIANA 46802

You may return this book to any agency, branch,
or bookmobile of the Allen County Public Library.

DEMCO

COMPUTER METHODS FOR TOLERANCE DESIGN

Robert G. Wilhelm
Rockwell International Corporation
Palo Alto, California, USA

Stephen C.-Y. Lu
Knowledge-Based Engineering Systems Research Laboratory
University of Illinois at Urbana-Champaign
Urbana, Illinois, USA

World Scientific
Singapore • New Jersey • London • Hong Kong

Published by

World Scientific Publishing Co. Pte. Ltd.
P O Box 128, Farrer Road, Singapore 9128
USA office: Suite 1B, 1060 Main Street, River Edge, NJ 07661
UK office: 73 Lynton Mead, Totteridge, London N20 8DH

COMPUTER METHODS FOR TOLERANCE DESIGN

Copyright © 1992 by World Scientific Publishing Co. Pte. Ltd.

All rights reserved. This book, or parts thereof, may not be reproduced in any form or by any means, electronic or mechanical, including photocopying, recording or any information storage and retrieval system now known or to be invented, without written permission from the Publisher.

ISBN 981-02-1058-2

Printed in Singapore.

Dedication

To my wife, Harshini
Robert G. Wilhelm

To my wife, Teresa, and my son, Derek
Stephen C-Y. Lu

Preface

Product development is a complex and multi-phase process requiring cross-disciplinary expertise from a team of engineers. To deliver products which simultaneously excel in high performance and quality with low cost and lead-time, specialized knowledge for design, analysis, and fabrication, must be synthesized, integrated, and applied across the whole spectrum of product life. This requirement leads to the concept of *concurrent engineering* which represents an improved product development methodology for shortening the time-to-market with highly competitive products. Realization of concurrent engineering is, however, not an easy task in engineering practice. It requires new corporate culture, organizational structure, and technological supports. To develop new technologies to support concurrent engineering, *integration* is the key challenge. For example, knowledge for various product life-cycle concerns must be integrated (*knowledge integration*); tasks across the facility, information, and decision levels must be integrated (*task integration*), and software tools for carrying out individual tasks must be integrated (*tool integration*). These integration requirements pose many research and development challenges that are being investigated by industrial and academic research organizations.

We are addressing these requirements with several broad research efforts at the Knowledge-Based Engineering Systems Research Laboratory (KBESRL) at the University of Illinois at Urbana-Champaign (UIUC). The *knowledge processing technology* (KPT) developed by KBESRL includes a set of software tools that increase the utility of engineering knowledge, including various know-how, know-why, and know-what, on computers (i.e., knowledge integration). These tools support complex engineering decision making tasks in an integrated and cooperative manner. In collaboration with its industrial sponsors, KBESRL's research results and KPT software tools have been applied to many real-life engineering problems. The current focus of KBESRL's research is to further develop KPT and demonstrate its applicability as a solution to concurrent engineering practice.

In our work with knowledge integration and concurrent engineering, we also see that advances in tolerance design serve the requirements of concurrent engineering. Tolerances and design margins are part of all engineering designs. Concurrent product and process design requires knowledge of the relationships between product function and process capability. The tolerances and margins expressed in individual designs are often the communication medium for negotiation during a concurrent engineering effort. The knowledge embedded in a tolerance specification is key to both *knowledge integration* and *task integration*.

In recent years there has been a significant increase in research and development [8, 70] devoted to tolerance analysis and synthesis. In part this effort follows advances in geometric modeling that now are commonly available in engineering design tools. As well, the computational resources now available allow for sophisticated and extensive tolerance calculations that might have been thought infeasible in the past. Most importantly tolerance studies have increased because industry demands quicker and more accurate results from both design and manufacturing engineers.

In this book we describe recent research advances and computer tools that can be applied in the determination of geometric tolerances. A framework for tolerance synthesis is developed and used with artificial intelligence techniques to provide computer methods for both analysis and synthesis of geometric tolerance specifications. While we discuss our work here in the context of geometric tolerances, we are also applying these techniques to many other kinds of design tolerances and margins. Our efforts in tolerance design address many of the integration issues associated with concurrent engineering.

This book comes after five years of active collaboration with many academic colleagues and industrial sponsors. We would especially like to thank our many sponsors for the generous support and encouragement that they have provided. In particular, the support of Rockwell International Science Center and Dr. Michael Buckley have helped us to continue collaboration between Palo Alto, California and Champaign, Illinois. We also appreciate very much the support of the Department of Mechanical and Industrial Engineering at UIUC.

Much of the software that we describe here was developed in collaboration with Allen Herman of KBESRL, Steven Brooks of Allied Signal Corporation, and Keith Hummel of Allied Signal Corporation (now with Grumman Corporation). As well, we acknowledge Dr. Placid Ferreira of UIUC, Dr. Terril Hurst of Hewlett-Packard Laboratories, Dr. Allen Ward of the University of Michigan, and Dr. Ramesh Srinivasan of the IBM T.J. Watson Research Center who have provided useful comments and suggestions. Thanks also to Amy Reinhart of KBESRL who has assisted during our research and book preparation. Finally, we thank all of the members of the Knowledge-Based Engineering Systems Research Laboratory who have contributed to our work through many interesting discussions of their related efforts.

This research was supported in part by the Digital Equipment Corporation, the U.S. Army Research Office under Contract DAAL 03-87-K-0006 with the University of Illinois Advanced Construction Technology Center, and by the National Science Foundation (DMC-86-57116).

Contents

Preface	**vii**
1 Introduction	**1**
1.1 Tolerances in Design and Manufacturing	2
1.2 Tolerance Analysis and Synthesis	7
1.2.1 A Typical Tolerance Synthesis Problem	7
1.2.2 A Typical Tolerance Analysis Problem	8
1.3 Unifying Synthesis and Analysis	9
2 Review of Related Research	**13**
2.1 Geometric Tolerances	13
2.1.1 Tolerance Analysis	13
2.1.2 Tolerance Representation	14
2.1.3 Tolerance Synthesis	15
2.2 Computational Tools and Approaches	17
2.2.1 Constraint-Based Reasoning	17
2.2.2 Feature Recognition and Explanation	18
2.3 Closure	19
3 Framework For Tolerance Synthesis	**21**
3.1 Current Practice	21
3.2 A Framework for Tolerance Synthesis	24
4 Primitives For Reasoning About Tolerances	**27**
4.1 Tolerance Theory	27
4.1.1 Virtual Boundary Requirements	28
4.2 Tolerance Primitives	32
4.2.1 Assembly Requirements and Fit	32
4.2.2 Empirical Requirements	40
4.3 Closure	45

5 Analysis Tasks and Sufficiency — 47
- 5.1 Tolerance Analysis — 47
- 5.2 Formulation for Sufficiency — 50
- 5.3 Inspection Procedure for Detection — 52
- 5.4 Closure — 54

6 Synthesis Tasks and Composition — 55
- 6.1 Composition and Competing Requirements — 55
- 6.2 Independent and Coupled Functional Requirements — 56
- 6.3 Approach for Composition — 58
- 6.4 Example for Independent Functional Requirements — 61
- 6.5 Example for Coupled Functional Requirements — 67
- 6.6 Closure — 70

7 Guiding Design and Explanation — 71
- 7.1 The Explanation Problem — 71
- 7.2 Structure for Explanation Task — 73
- 7.3 Rule-Based Approach — 76
- 7.4 Extensions for Optimization Formulations — 80
- 7.5 Closure — 82

8 CASCADE-T Implementation — 85
- 8.1 Design Scenario — 85
- 8.2 Description of Implementation — 86
- 8.3 Examples — 91
- 8.4 Closure — 102

9 Conclusions — 103
- 9.1 Computer Methods for Tolerance Design — 103
- 9.2 Open Areas for Research — 106
  - 9.2.1 Tolerance Representation — 106
  - 9.2.2 Sufficiency — 106
  - 9.2.3 Validity and Composition — 106
  - 9.2.4 Explanation — 107
  - 9.2.5 Extending CAD Tools — 108

Appendix — 109

A Tolerance Primitive Definitions — 109

B Tolerance Sufficiency Examples — 119

C Constraint Explanation Examples	**125**
Bibliography	**135**
Index	**142**

List of Tables

1.1	Tolerance Analysis Example	9
4.1	Parametric Space for Cylindrical Solid	34
4.2	Example of Cylindrical Fit Primitive	35
4.3	Possible States for a Necessary Condition	36
4.4	Example of Several Necessary Conditions	38
4.5	Example of Confounded Parameters	39
4.6	H6h6 ISO Tolerances	41
4.7	Operating Ranges for Walsh's Bearing Design Method	42
4.8	Best Engineering Practice for Common Journal Bearings	43
4.9	Example of Bearing Design Primitive	46
5.1	Tolerances Assigned in Tolerance Analysis Example	49
6.1	Starting Values for Prisms	62
6.2	Determined Values for Prisms	62
6.3	Trace From Initial Prism Calculations	64
6.4	Multiplication Constraint Definition	65
6.5	Modified Values for Prisms	65
6.6	Trace From Modification of Prism Parameters	66
7.1	Trace of Prism Example Constraint Explanation	73
7.2	Partial Trace of Tolerance Primitive Constraint Explanation	74
7.3	Lisp Functions for Why Explanation	77
7.4	Lisp Functions for How Explanation	78
7.5	IDEEA Database Facts for Prism Explanation	79
7.6	Aggregated Explanation for Prism Explanation	80
7.7	IDEEA Rules for Prism Explanation	81
7.8	Intermediate Results for Optimization Formulation	83
8.1	Open a Model File in CASCADE-T	88
8.2	Example Constraints for Model File in CASCADE-T	90
8.3	Tolerance Primitives for Assembly Requirements	98
B.1	Review of Geometric Constraints	120

B.2	Line Definitions for Tolerance Analysis Example	121
B.3	Geometric Tolerances for Tolerance Analysis Example	122
B.4	Dimensions and Default Tolerances for Tolerance Analysis Example .	122
C.1	Trace of Tolerance Primitive Constraint Explanation (1 of 7)	126
C.2	Trace of Tolerance Primitive Constraint Explanation (2 of 7)	127
C.3	Trace of Tolerance Primitive Constraint Explanation (3 of 7)	128
C.4	Trace of Tolerance Primitive Constraint Explanation (4 of 7)	129
C.5	Trace of Tolerance Primitive Constraint Explanation (5 of 7)	130
C.6	Trace of Tolerance Primitive Constraint Explanation (6 of 7)	131
C.7	Trace of Tolerance Primitive Constraint Explanation (7 of 7)	132

List of Figures

1.1	Geometric Tolerances in the Enterprise	3
1.2	Tolerance Synthesis Example	8
1.3	Tolerance Analysis Example	10
3.1	Tolerance Errors	22
3.2	A Framework for Tolerance Synthesis	25
4.1	Virtual Surfaces for Pin and Washer Assembly	29
4.2	Virtual Surfaces for Material Bulk of Tube	29
4.3	Parameterized Stud for Stud/Hole Pair	31
4.4	Tolerance Primitives for Assembly and Fit	33
4.5	Bounding Approximation for Tolerance Primitive	41
4.6	Minimum Recommended Diameters for Journal Bearings, From Walsh	42
5.1	Geometric Constraints for Tolerance Analysis Example	48
6.1	Composition of Independent Primitives	57
6.2	Composition of Primitives with Coupling	59
6.3	Independent Functional Requirements Example	61
6.4	Constraint Network for Prisms	62
6.5	Coupled Functional Requirements Example	68
7.1	Hierarchical Structure for Explanation Task	75
8.1	BRL Display for Model File in CASCADE-T	89
8.2	Typical CASCADE-T Display	90
8.3	Bearing Example, Initial Problem	91
8.4	Bearing Example, Searching Possibilities	92
8.5	Bearing Example, First Solution Infeasible	93
8.6	Bearing Example, Feasible Solution	93
8.7	Bearing Example, Changing Tolerances	94
8.8	Bearing Example, Larger Bearing and Tolerances	95
8.9	Assembly of Base, Shaft, and Plate	96
8.10	Labels and Dimensions for Assembly of Base, Shaft, and Plate	97
8.11	Variation of the Solid Model From CASCADE-T	99

8.12	Composition of Assembly Requirements in CASCADE-T	99
8.13	High-Level Constraint Network Graph for Assembly	100
8.14	Detailed Constraint Network Graph for Assembly	101
B.1	Tolerance Analysis Example	121
B.2	Variable Labels for Tolerance Analysis Example	123
B.3	Partial Graph for Tolerance Analysis Example	124
C.1	Tolerance Primitive Constraint Graph	133

Chapter 1

Introduction

Tolerances and design margins are applied in all engineering decisions. For example, when knowledge of physical principles is limited, design margins ensure acceptable performance for wide variations in physical behavior. Similarly, when cost effective variations are determined for manufacturing processes, geometric tolerances ensure that slight variations in shape all provide acceptable performance. Tolerances and design margins accommodate the uncertainty that is inherent in engineering practice.

With the introduction of concurrent engineering techniques, tolerances take on an even larger role in engineering documents and descriptions. Concurrent product and process design requires knowledge of the relationships between product function and process capability. The tolerances and margins expressed in individual designs are often the communication medium for negotiation during a concurrent engineering effort. In this book, we present several new techniques for tolerance design that allow for integration among many engineering disciplines. While the discussion and examples are generally posed with regard to geometric tolerances, these techniques are applicable to all types of design margins and tolerances.

A geometric tolerance describes the degree to which a nominal design feature can vary while satisfying functional requirements. For example, a lever transmits force when the applied force exceeds equilibrium conditions and the lever is sized to carry the transmitted force. If a designer requires a lever that transmits forces in the range of one to ten pounds, the minimum lever cross-section will often be determined by the worst-case loading condition of this design requirement. All lever cross-sections that exceed this minimum will satisfy the design requirement. If the designer chooses the minimum cross-section as nominal, larger cross-sections would be acceptable since they satisfy the design requirement. In this example, then, the area of the cross-section is a geometric feature and the force range is a functional requirement. The tolerance on the nominal feature would -0 and $+\infty$.

A geometric tolerance also describes the variation that is allowed when fabricating parts from a design. Manufacturing processes have particular and repeatable ranges of variation. The tolerance associated with a process depends on expected process variation and the techniques used for measurement. It is most cost effective to choose geometric tolerances and manufacturing processes so that allowable design variations are larger than expected manufacturing variations. For the lever example, the positive tolerance on the nominal feature, $+\infty$, would be reduced once a fabrication process was selected.

Using slightly different interpretations, designers and manufacturing engineers use the same geometric features and tolerances to establish designs and fabrication plans. Geometric tolerances are one class of engineering design results where details influence concepts and small changes may affect both design and manufacturing requirements. For example, when standard tooling exists in a manufacturing facility, design modifications calling for small changes in dimensions or tolerances may have a high cost while large changes consistent with existing tooling may be very inexpensive to implement. Further, while tolerances are applied time and time again, each new design introduces variations which may require that tolerance specifications be reconsidered.

1.1 Tolerances in Design and Manufacturing

To illustrate how geometric tolerances are related to many aspects of design and manufacturing, we consider the dataflow diagram of Figure 1.1 which suggests how design and manufacturing data are manipulated during the design and fabrication stages of a product lifecycle. Note that, in the figure, rounded boxes suggest computed values while square boxes suggest computing tasks. Double-lined boxes suggest where results in design and manufacturing are achieved.

Initially, as a design is developed, functional requirements and nominal geometry are determined. A functional requirement is a statement of performance that is required for the product under design. The load characteristic of the lever discussed earlier is a very simple example of a functional requirement. Products that are made from an assembly of components and subsystems may include many different levels of requirements. For example, in the broadest sense, the functional requirements imposed for the design of a compact disk player are to reliably provide sound reproduction of high fidelity while requiring a limited amount of labor and material for fabrication and use of the equipment. Each subsystem in the product, in turn, has more detailed requirements that contribute to this overall performance of the product. For example, a typical compact disk player makes use of a low friction suspension to support and direct the reader head as it travels radially along the compact disk below. This suspension system must be designed to provide the appropriate damping as the head travels and to minimize the inertia that must be overcome when the head motion is initiated. Further the variation in form and size allowed for the parts in the suspension must be specified to coincide with the friction characteristics while preventing inaccurate and unexpected motion.

Nominal solids are the geometric forms shown in engineering drawings to specify components or subassemblies of a product under design. Topology refers to how the

1.1. TOLERANCES IN DESIGN AND MANUFACTURING

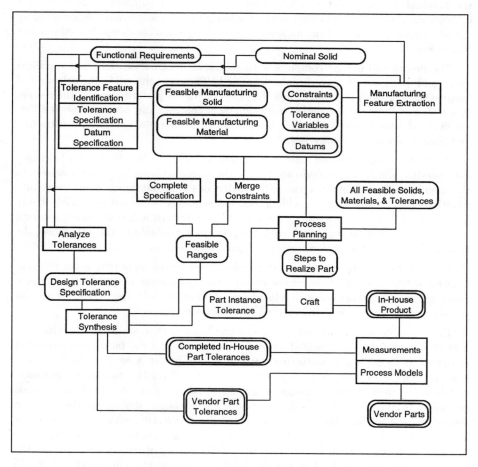

Figure 1.1: Geometric Tolerances in the Enterprise

geometric entities of a component are connected or arranged. In practice, the topology specified in the drawing will coincide with the topology of parts that are fabricated. Geometry refers to the actual size and position of geometric entities. The dimensions and geometry of actual parts will differ slightly from the dimensions specified in the nominal drawing. While it is convenient to use real numbers to describe part dimensions, in practice each dimension is achieved within a tolerance determined by the fabrication process and the measurement technique that is chosen to evaluate the dimension.

The design will not be complete until tolerance information, material specifications, and other details have been determined. These details guide the selection of fabrication strategies and also define the range of robust performance for the design. In arriving at these details, design engineers must consider and develop several different kinds of interrelated data which include datums, tolerance variables, constraints, and feasible shapes and materials.

A feasible manufacturing solid is a variational model, described by a tolerance specification, which can be fabricated using the manufacturing processes available to the enterprise. For example, a part with cavities of significant depth would constitute a feasible solid for a casting or molding shop but probably would not be feasible if only drawing and die-press processes were available. Similarly, a feasible manufacturing material is one that is consistent with the processes available to the enterprise, the form required, and the performance specified for the design.

In the tolerance representation, constraints refer to the equations and rules that determine how functional requirements and geometric form are related. In many cases, geometric entities are also related by constraints. Tolerance variables are the variations that are allowed or expected for each geometric entity.

Datum surfaces and features are geometric entities that are used to arrive at a best fit between design or manufacturing descriptions and the fabricated form that is arrived at through manufacturing steps. For example, it is common to define the position of a line or plane according to points positioned relative to a three dimension coordinate system of three orthogonal axes. In practice, strict orthogonality cannot be achieved and an approximation is used. Three planes, defined as the primary, secondary, and tertiary datums, define the coordinate system that is used to measure the position of geometric features in a design. Measurements are made relative to planes that are "fitted" to the datum surfaces. The order of the datums specifies the priority of each datum. That is, the fit of the primary datum is undertaken first and its determination then influences the fit of the secondary datum and so on.

A tolerance specification is determined by describing the variation allowed for each geometric entity that is used to describe the geometric form of a part. This is achieved by identifying the geometric features, datums, and the relationships that exist among datums and geometric features such that particular functional requirements are satisfied.

1.1. TOLERANCES IN DESIGN AND MANUFACTURING

Tolerance feature identification refers to the activity associated with pairing functional requirements and geometry. In practice, only some portions of a design are assigned tolerances. Ideally, allowable variation should be specified for all geometry. Because some design requirements interact, geometric variations may be determined from more than one requirement. For these cases, it is important to identify the ways in which particular functional requirements influence the allowed geometric variation.

Datum specification refers to the activity of identifying useful datums for design and manufacturing descriptions. The determination of datums is related to particular parts and the current engineering task. During design, it is often useful to define datums according to part interaction. During manufacturing planning, it is more useful to define datums according to the geometric constraints imposed by fixturing and material processing.

The combination of all nominal solid descriptions, material, and tolerances describes a class of different parts that may be fabricated. To determine the best approach for manufacturing within an enterprise, it is useful to consider the subset of this class of parts that can be manufactured using the processes available to the enterprise. Often this subset of feasible combinations can be determined by matching standard features with the design description.

Manufacturing feature extraction refers to this activity of associating parts of the design description with manufacturing operations that can be performed. For example, each surface that describes the envelope of a part might be achieved by cutting material away from a blank. In this case, the manufacturing features are cutting procedures that correspond to enveloping surfaces. In another case, the features may correspond to the cavities that can be casted or molded.

Feasible solids and material as well as datums, constraints, and tolerance variables are all part of the tolerance specification that is shared by design and manufacturing engineers. Working with this specification, additional computing must be done to ensure completeness and consistency.

A complete tolerance specification implies that all geometric variations of the product are bounded. Further, the bounds are derived from a set of consistent functional requirements. In effect, the tolerances specified for a design describe all product attributes that are negotiated between design and manufacturing concerns. As changes are made to the tolerance specification, it is useful to measure the completeness of the specification. Once completeness is reached, there is no necessity to add further tolerances to the design unless the functional requirements change. In many designs, more than one functional requirement may influence each geometric entity. These competing requirements must be resolved to yield a tolerance specification that satisfies all design requirements. Merge constraints refers to this resolution activity. Note that complete tolerance specification may include conditional tolerance relationships. That is, one design or tolerance parameter may be specified as a function of another. Feasible ranges refers to the many combinations of allowable tolerances that result from these conditional tolerances.

Given a complete and consistent tolerance specification, further computing is then done to choose the final tolerances that are specified for the drawing. Analysis and synthesis tasks are both needed to determine the final tolerances.

As a design is completed, it is common, in current practice, to determine if the tolerances specified for a design are consistent with design requirements. Tolerance analysis refers to this determination where a tolerance specification and design requirements are input and calculations are performed to determine if the variational class defined by the tolerance specification is within the class required by the design requirements.

As fabrication strategies are considered and selected, the feasible ranges in the design are reduced to a final design tolerance specification. As well, in assemblies of parts, tolerances are distributed over design features to minimize manufacturing costs.

The design tolerance specification is arrived at through tolerance synthesis. This task is analogous to an optimization procedure. The objective function is most often characterized according to cost and the constraints that describe the feasible region describe functional requirements, geometric relationships, and manufacturing capabilities.

In an assembly of parts, each distinct part is usually manufactured separately. Part instance tolerances describe the allowable variation for each distinct part in the design. Manufacturing plans are then developed to satisfy the shape, material and tolerance requirements of each distinct part. When the design and manufacturing are both performed within a single enterprise, part instance tolerances are often a subject of negotiation among design and manufacturing engineers.

Prior to manufacture, a description of manufacturing procedures is determined for each part. This process plan describes the manufacturing processes, parameters, and sequences that are necessary. These are the steps that must be taken to realize the part. Following the process plan, manufacturing craftspeople use machines and processes to complete each part in the design. Often knowledge and feedback from craftspeople also influences the tolerances selected for particular part instances.

The result is an in-house product that reflects the processes and variation of the manufacturing enterprise. Note that for novel designs, the requirements expressed in a tolerance specification will determine the manufacturing processes and operating ranges that must be developed.

Measurements are made during and after manufacture to monitor process variability and ensure that manufacturing results coincide with design specifications. When process monitoring is not feasible, mathematical models may be used to estimate the process behavior that yields particular part measurements. Here the variation measured in manufactured parts predicts the state of the manufacturing process. For new processes, these process models are also used to select appropriate operating parameters for process plans. As a manufacturing enterprise develops experience with particular manufacturing process and materials, a database of measurements for completed parts can be assembled. This database describes the class of tolerances that can be achieved for parts manufactured in-house.

Not all parts in a design are manufactured within a single manufacturing enterprise. Vendor parts, supplied according to specifications, are also used. Commonly, as vendor parts are received, some measurements are made to ensure that design requirements will be met. While manufacturing processes are stable, these measurements serve as a guide for tolerance synthesis. When processes degrade or improve, the new measured tolerances serve as additional constraints for tolerance synthesis.

1.2 Tolerance Analysis and Synthesis

Figure 1.1 illustrates how tolerances are part of many design and manufacturing activities. It should be clear that computational tools for tolerance design can influence early and late stages of the design lifecycle as well as many different aspect of engineering practice. We will now take a closer look at the different computing techniques that are employed for tolerance design.

Design and computing techniques for geometric tolerances are usually distinguished as either analysis or synthesis. Tolerance synthesis takes part geometry and functional requirements as input and results in tolerance specifications. Tolerance analysis requires a tolerance specification and design requirements as input and determines if the variational class defined by the tolerance specification is within the class required by the design requirements.

1.2.1 A Typical Tolerance Synthesis Problem

A typical tolerance synthesis problem is illustrated in Figure 1.2. The cylindrical stud on the left side of the figure must assemble within two mating parts with holes as shown on the right side of the figure. A complete tolerance specification for this assembly would describe the variation allowed for each feature in each part of the assembly. The synthesis problem is to choose allowable variations for each feature so that functional requirements such as assembly are satisfied. Further, cost efficient tolerances should be selected so that manufacturing expense is minimized. The geometric tolerances on Figure 1.2 indicate that the nominal diameter of the smaller cylindrical feature is 1.0 inch. The orientation of the cylindrical feature is defined relative to datum A. For the smaller cylindrical feature, tolerance synthesis involves choosing appropriate values for variation in diameter and perpendicularity as indicated by the question marks. Similar selections must also be made for the other features in the assembly.

Tolerance synthesis is difficult because there are many feasible solutions to each set of requirements and because, in general, necessary and sufficient conditions for conforming specifications are not known. To date, most efforts to develop computer-

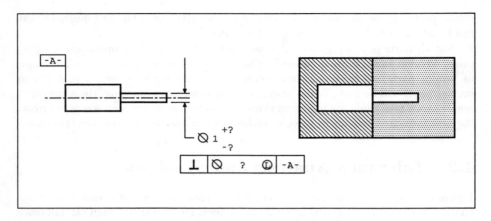

Figure 1.2: Tolerance Synthesis Example

based tools for tolerance synthesis have involved optimization techniques applied to particular design problems.

1.2.2 A Typical Tolerance Analysis Problem

A typical tolerance analysis problem is illustrated in Figure 1.3. Each of the dimensions shown in the figure has been assigned a default tolerance of ±0.01. The variation of actual dimensions is assumed to be normally distributed. Expecting that dimension **Length-1** significantly influences the performance of an assembly, a tolerance analysis is performed as shown in Table 1.1. The analysis suggests that the current tolerancing scheme will produce a worst-case variation in **Length-1** of ±0.03. The probability in tolerance of 90.4% indicates the percentage of parts that will meet the ±0.01 tolerance specified for **Length-1**. The table of contributors indicates the degree to which tolerances of related dimensions contribute to the variation in **Length-1**. The sensitivity column indicates how a change in each related dimension influences **Length-1**. In this case, for dimensions e, c, and d, a one unit change causes a one unit change in **Length-1**. The % Contrib. column indicates that each of these dimensions contributes at the same rate. The last row of the table of contributors refers to a parallelism requirement specified for the left and right sides of the part.

Tolerance analysis results help engineers to infer changes in design and tolerance parameters that are necessary to produce a feasible specification. Tolerance analysis is difficult because appropriate analysis conditions are not always apparent. For example, the best change may involve design features that are not part of the current tolerance specification. Further, many analysis cases may have to be considered for tolerance specifications where less than full acceptance is required. To date, computer-based tools for tolerance analysis provide sensitivity analysis with respect to perfect

Table 1.1: Tolerance Analysis Example

Mechanical Advantage Tolerance Analysis

Name	Length-1
Value	4.0000 in
Probability in tolerance	90.4%
Derived Tolerance	-.01803 (3.9820) / 0.1830(4.0180)
Upper tolerance	+.01000 (4.0100)
Lower tolerance	-.01000 (3.9900)
Dimensional Worst Upper	+.03000 (4.0300)
Dimensional Worst Lower	-.03000 (3.9700)

#	Name	% Contrib	Upper	Lower	Type	Sensit.	Value	Upper	Lower
1	e	30.8%	+.010	-.01000	D	-1.00	1.00 in	+.010	-.010
2	c	30.8%	+.010	-.01000	D	+1.00	6.00 in	+.010	-.010
3	d	30.8%	+.010	-.01000	D	-1.00	1.00 in	+.010	-.010
4	p-1	7.7%	+.005	-.00500		.50	—	.010	—

form shapes and simulation tools to test acceptance ratios for assemblies with statistically distributed dimensions. These tools are quite restricted in representation capabilities and have very limited and rigid query facilities.

1.3 Unifying Synthesis and Analysis

New approaches to computing make it possible to unify many aspects of tolerance analysis and synthesis procedures. In this book, we describe methods for this unification that employ geometric tolerance specifications for assemblies of parts. Tolerance primitives, based on a sound theory of tolerancing, are used to represent tolerance relationships or links between geometric entities and functional requirements. Algorithms are developed for the determination of boundedness and the measurement of sufficiency. A detailed constraint network is used to represent tolerance relations for a part under design and provide for the composition of tolerance specifications. The constraint network is, in turn, associated with nominal solids describing the shape of the part under design. Query methods for the constraint network support the measurement of tolerance sufficiency and tolerance synthesis through optimization. The representation and computational methods are developed within a framework for tolerance synthesis which includes seven interconnected computing tasks. To demonstrate the improved facilities for tolerance specification which result from the integration and interaction within the framework, a prototype computing environ-

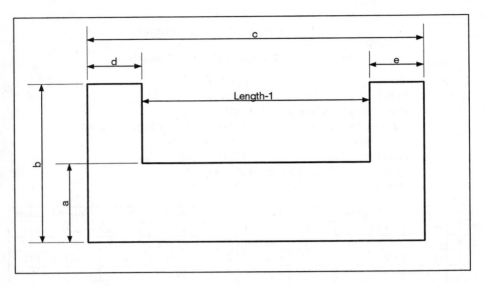

Figure 1.3: Tolerance Analysis Example

ment has been developed. The computing environment is called **CASCADE-T** : **C**oncurrent, computer-**A**utomated methods for the **S**ynthesis of **C**ompeting **D**esign **E**lements and geometric **T**olerances. Synthesis is emphasized because CASCADE-T allows tolerance specifications to be generated and checked automatically during the design process.

While this work should be immediately applicable in many engineering tasks, the methods and implementation described here are even better suited for the demands of concurrent engineering. The concurrent engineering paradigm suggests that significant improvements in design and manufacturing productivity can be achieved by considering design and manufacturing concerns at the same time. Advances in computer-aided tolerancing are important to this paradigm since concurrent product and process design requires knowledge of the relationships between product function and process capability and a complete tolerance specification describes these relationships.

This work provides improved methods for early design checking to ensure that both performance and manufacturing requirements are satisfied. As well, the detailed representation developed in this work may be viewed from many different engineering perspectives to allow for the optimization of both product and process requirements. Though demonstrated for geometric tolerancing, this research is applicable to many types of engineering design where knowledge and features from past designs can be modified and reused.

1.3. UNIFYING SYNTHESIS AND ANALYSIS

The remainder of the book is presented in three parts. The motivation and basis of the research are presented first, in Chapters 2 and 3, with a review of related work and a presentation of the framework for tolerance synthesis. The techniques developed for analysis and synthesis are then presented in Chapters 4 through 7. Representation and computing techniques are described for tolerance primitives, sufficiency, composition, and explanation. Finally, an implementation and the results of the work are discussed in Chapters 8 and 9 along with conclusions and a discussion of future research.

Chapter 2

Review of Related Research

Research in the computation of geometric tolerances is generally posed as either an analysis, synthesis, or representation problem. The earliest work focused on the manipulation of dimensions and the analysis of tolerance specifications. Later approaches have accommodated dimensions, representation of tolerance information, techniques for testing feasibility, and synthesis for prespecified geometric models. More recent work has focused primarily on the underlying theory of tolerancing, composition, and properties such as validity and sufficiency. Previous work in these three areas is discussed in detail.

The approach to computation taken in this book is significantly different from past work in geometric tolerances. Very detailed constraint networks are used to represent and manipulate tolerance relationships. Feature recognition is required to provide useful explanations of the detailed constraints. Other efforts in computer science and geometric reasoning have made use of the same computational approaches. A brief discussion is provided for both constraint-based computation and feature-based recognition. To conclude the chapter, the techniques used in this book are compared to previous efforts.

2.1 Geometric Tolerances

2.1.1 Tolerance Analysis

Tolerance analysis requires a tolerance specification and design requirements as input and determines if the variational class defined by the tolerance specification is within the class required by the design requirements. Fortini [17] summarizes analyses commonly used to test worst-case conditions or to meet a particular statistical criteria. Hoffman [26] shows techniques for calculating nominal, minimum, and maximum part dimensions and discusses how tolerances might be allocated among several process steps. Fainguelernt, Weill, and Bourdet [13] describe analysis techniques that account for manufacturing capabilities. Greenwood [21] introduces statistical models which also account for process characteristics when estimating tolerance build-up. A detailed treatment of tolerance chains, statistical methods, and tolerance allocation is given by Bjørke [4]. Turner reviews capabilities in current commercial software [74] for tolerance analysis.

Working with tolerances or dimensions specified on perfect-form shapes, Hillyard [25] shows how admissible dimensioning can be detected and the sensitivity of tolerances to different dimensions can be calculated. Tolerance sensitivity analysis relates changes in tolerances to changes in dimensions. A graph theoretic approach to dimension consistency in one dimension, first discussed by Requicha, was extended by Todd [71] to two dimensional cases. Geometric entities with more than two constraints are overdimensioned while entities with fewer than two constraints are not defined adequately for manufacture.

Constraint networks have been suggested by Fleming [16] as a possible tool for tolerance analysis tasks. Different constraints for datum-zone, datum-datum, zone-zone, and zone-datum combinations are proposed and methods are suggested so that a constraint graph might be queried to perform typical tolerance analysis tasks.

2.1.2 Tolerance Representation

In many computer-aided design (CAD) systems it is common to preserve some information about dimensions and tolerances. Gossard [20] describes dimension-driven geometry that can be used to incrementally refine a part shape as dimensions are established. This approach works from a perfect form characterization of the part under design and has limited support for tolerance information. Most notable, dimension-driven systems allow incremental refinement and enforce strong admissibility criteria.

Requicha [55] describes extensions to the PADL solid modeler that support information and relationships that describe tolerance specifications. This work demonstrates the utility of maintaining consistency among tolerance variables and solid model attributes.

Many improvements and refinements have been suggested for the representation of tolerance specifications. Ranyak and Fridshal [52] have developed a hierarchical approach to feature modeling and a dimension and tolerance model for representing tolerance specifications. Their implementation can represent many of the ANSI tolerances and has been demonstrated with a process planning application. Roy and Lui [56] propose a hybrid representation that combines constructive solid geometry, boundary representation and structured face-adjacency graphs. Their model can be used to provide multiple levels of abstraction and to support reasoning about calculation dependencies. Etesami [12] describes a method for including manufacturing information within the part model. Each part feature is linked with a bounding solid constructor. These constructors and their accompanying information can then be used to verify that manufactured parts meet tolerance specifications. With further development, it is suggested that this approach will automate inspection procedures

2.1. GEOMETRIC TOLERANCES

used for verification. Shah and Miller [58] discuss the functional requirements of representing tolerance specifications and describe an implementation that enhances the system developed by Ranyak [52]. Each of these systems seeks to provide adequate support for the ANSI standard and provide information that would be appropriate for tolerance analysis - no direct analysis capabilities are provided in these systems.

It is common, in practice, to employ specially formulated representations when determining tolerance. Industrial standards [1] describe many special cases from which tolerances can be determined and reflect common gauging practices. With the introduction of coordinate measuring machines, more flexible methods for tolerancing and measurement are now practical. Lehtihet and Gunesena [34] have developed algorithms for the gauging of patterns of holes. Algorithms for flatness and straightness are presented by Traband, et al. [72].

Two theories of tolerancing, recently proposed, provide some basis for the formulation of general tolerancing models. Requicha discusses the issues behind such a general approach in [54] and proposes a theory of geometric tolerancing in [53]. Geometric constraints on two dimensional subsets of objects define a tolerance specification. While this work represents a major step toward a theory of tolerancing, it also demonstrates the difficulties that must be overcome before form and position tolerances can be handled accurately in a general way. Working with a smaller range of functional requirements, Jayaraman and Srinivasan [27, 62] have proposed the Virtual Boundary Requirements (VBR) method for representing tolerances associated with assembly and material bulk requirements. Surfaces associated with design features and fixed relative to each other are used with tolerance assertions to describe tolerance zones. For these functional requirements, they show properties and develop theorems that allow conditional tolerance relationships to be derived from VBR specifications.

Several other studies have been made of representation schemes for assemblies of parts. Wang and Oszoy [78] have developed a representation that supports the automatic generation of tolerance chains. Srikanth and Turner [60] have proposed a more general representation that preserves information about the dependencies that exist between parts in an assembly. A similar approach for dynamic and kinematic analysis is described by Kim and Lee [29]. Treacy, Ochs, and Oszoy [73] show how spatial relationships can be used to represent an assembly of parts and to calculate confidence limits for dimensions in the assembly.

2.1.3 Tolerance Synthesis

Tolerance synthesis takes part geometry and functional requirements as input and results in tolerance specifications. Working with a part specification and dimensions, Turner [75] describes a theory of tolerance modeling that has been implemented with a CAD system to choose the best tolerances for particular dimensions. In this approach, linear approximations are employed along with linear optimization techniques. The focus here is on modeling methods that are reasonably accurate and computationally tractable. One variable optimization and three dimensional sensitivity analysis are supported. Another common approach for tolerance synthesis makes use of Monte Carlo simulation. Lehtihet and Dindelli [35] describe one such simulation tool. Several others are now available from commercial vendors.

A number of other studies have been made of tolerance selection and synthesis problems. Kulkarni [30] suggests an estimate that cost of production will vary inversely with the square of the tolerance. Using this approximation, a LaGrange Multipliers solution is developed for allocation of tolerances to component dimensions subject to constraints on the final assembly tolerance.

Lee [31] shows methods for optimum selection of standard tolerances for dimensions according to cost criteria and constraints for stack-up considerations. This approach exploits the limited number of discrete tolerance ranges specified for standard fits and uses a branch-and-bound algorithm to efficiently consider alternatives. This papers also makes strong arguments for monotonic relationships between cost and tolerances. A rule-based approach to this kind of selection is described in [41].

Parkinson [48] uses a reliability model to calculate risk values associated with tolerances chosen for features in a design. The risk values are computed according to limit state equations derived from the design specification. It is suggested that when cost functions are known, the risk values can be used to optimize tolerance selection according to cost. Lee [32] describes a nonlinear optimization approach for tolerance synthesis that uses a reliability index and guarantees convergence. Numerical optimization techniques are commonly applied in a wide variety of engineering domains [76]. The addition of tolerancing variables to the general design optimization problem is discussed by Michael in [42] and [43]. Additional cost minimizations are discussed by He [22]. Allocation methods which include manufacturing process selection are presented by Chase, Greenwood, and Losli [6].

The manufacturing cost associated with particular tolerances is most often minimized during tolerance synthesis. Models for estimating the cost of dimensional tolerances are described by Ostwald and Blake [47]. A comparison of cost-tolerance functions and their use in analysis and allocation is presented by Wu, Elmaraghy and Elmaraghy [87]. Using an expert system that assigns appropriate cost models for each design tolerance, Dong and Soom [11] describe an optimization approach that can be applied to multiple and related design chains.

The objective function and constraints used for tolerance synthesis are directly related to the function or performance associated with the part under design. Weill [80] and Farmer [14] review the requirements for a functional approach to tolerancing. Functional models proposed for general design are also discussed by Schmekel [57].

2.2 Computational Tools and Approaches

2.2.1 Constraint-Based Reasoning

Constraint-based reasoning has been studied in a variety of ways. A brief account is provided by Ward in [79]. *Constraint satisfaction* tests the feasibility of a solution by efficiently testing constraints that must hold for the current problem. *Constraint directed search* dynamically prunes search space by locally applying constraints to the current solution. *Constraint languages* seek to represent mathematical relationships so that unknowns may be automatically inferred from knowns and contradictory relationships are quickly identified. *Constraint compilation* merges constraints describing requirements and available properties to yield sets of feasible designs. *Constraint abstraction* encapsulates constraints and resultant designs to maintain design rationale. The following review focuses on one engineering application in each of these different approaches to constraint-based reasoning.

The ROSCAT computer-aided tolerancing system developed by Manivannan [41] provides a recent example where constraint satisfaction is used. Sutherland's SKETCHPAD [68] was perhaps the first computer program to use this approach. In ROSCAT, constraints describe geometric relations that must hold for each type of fit. The user provides geometric information about mating components and the ROSCAT suggests standard fits.

The ISIS-II job-shop scheduling system described by Fox [18] provides a very good example of constraint-directed search. Constraints are used to describe production goals, physical capacity, sequencing requirements and preferences. These constraints are then used to guide the search for production schedules.

The CONSTRAINTS language developed by Steele [67] and other subsequent implementations by Herman [24] and Simmons [59] demonstrate the capabilities of constraint languages. In CONSTRAINTS, mathematical relationships are represented as constraints. A user may connect instances of these constraints in a network to describe a design problem. Dependency information is maintained as values are set in the network of constraints so that calculations can be explained and revised correctly. There are no explicit input/output relationships - calculations flow from known values to unknown values. As values are entered for particular design parameters, other design parameters are calculated by propagating these known values through the constraint network. CONSTRAINTS also provides for the construction of compound constraints from groups of primitives. In the electronic circuit problems that CONSTRAINTS was applied to, strictly hierarchical networks were not sufficient to represent the multiple views employed during efficient analysis and design cycles. Carefully employed *Almost-Hierarchical* relations were used in CONSTRAINTS to improve the representation.

The Pride system for paper handlers described by Araya [2] and Ward's system for power transmission systems [79] are compilers of constraints. Functional, structural and performance requirements are represented with constraints which, differing from general constraint languages, have explicit inputs and outputs. These constraints are then compiled into a design plan which describes a collection of designs that would satisfy the requirements.

The design rationale system described by Thompson [69] provides for additional abstraction of constraints. Constraints represent design rationale, or the reasons behind the resulting design, and may be of several types : mathematical, procedural, goal/plan, set, rule or text. As a design is carried out, design activity is captured using these different constraints.

2.2.2 Feature Recognition and Explanation

Feature recognition systems assume that some set of primitive features have been predefined. The problem is then to recognize patterns of primitive features that require particular machining, manipulation, or interpretation. Much work has been done to use feature recognition techniques to identify manufacturing features in solid models and then automatically generate manufacturing plans. Choi [7] developed a system which extracted manufacturing features directly from a CAD representation. Staley [63] describes a classification method and syntactic pattern recognition methods which can be used to test for realizability and select processes. Li [36] extended this approach with a computer system that is useful for representing turned parts and recognizing manufacturing features. Srinivasan [61] describes a system for selecting machine and tool combinations which can generate the surfaces of a workpiece. Joshi [28] presents a feature representation scheme, based on boundary representation and attributed adjacency graphs. Features are defined by unique relationships between boundary-representation faces. A graph-matching heuristic is then used to match features with subgraphs on a graph which represents a part.

These systems all build on similar efforts in picture interpretation. Mackworth [40] summarizes early work in picture interpretation along with some reasons for difficulty. Namely, the theories employed for interpretation are incomplete. Sufficient conditions abound but complete necessary conditions are not to be found.

The feature recognition task is very similar to the procedure required to interpret or explain a system of constraints. The problem of explaining constraint networks was addressed by Sussman and Steele [67] in their CONSTRAINTS language. In their approach, aggregation is handled with multiple levels of constraints. Their method for generating explanations uses truth maintenance information and employs several methods for pruning the explanations. In some respects, interpreting constraints is quite similar to one addressed by Lee and Fu [33] concerning recognition of constructive solid geometry (CSG) features. In facing this problem, Lee and Fu use domain knowledge at increasing levels of detail to prune the possible explanations. A much more detailed approach to domain specific explanation can be seen in the RAPT interpreter [49] described by Popplestone et al. In this system assembly plans are arrived at by determining (or explaining) a sequence of relations that will connect a goal to an initial state. The concept of aggregation and detailing is also explored by De Floriani and Falcidieno [10] in their work on hierarchical boundary models. Additional features are reduced to boundary representation and then manipulated using refinement and abstraction operators.

2.3 Closure

Engineering approaches to tolerancing are commonly biased toward either design or manufacturing concerns. The associated tolerances of a design might be optimized to meet design objectives, or the process variation of a manufacturing process might be used to bound designs under consideration. As well, most efforts have been directed at either analysis or synthesis but not both. The concurrent engineering paradigm [37] suggests that significant improvements in design and manufacturing productivity can be achieved by considering design and manufacturing concerns at the same time. Advances in computer-aided tolerancing are important to this paradigm since concurrent product and process design requires knowledge of the relationships between product function and process capability and a complete tolerance specification describes these relationships [39].

Weill [81] and Farmer [14] suggest that a functional approach to tolerancing is required to provide this capability. In this book, we describe the computational tools that are required for a functional approach to tolerancing. Working from the tolerance representation proposed by Jayaraman and Srinivasan [27, 62], tolerance primitives have been developed to support representation and computation that is focused on functional requirements.

To consider design and manufacturing concerns at the same time, techniques must be available for the composition of design and manufacturing features. The techniques described in this book extend the approach, described by Requicha [53], of composite features and provides a very general mechanism for merging design and manufacturing requirements. Further, for particular classes of requirements, the VBR approach [27, 62] is employed to define necessary and sufficient conditions for acceptable tolerances that are quite useful for tolerance synthesis.

Design synthesis among several engineers of different disciplines requires that varying but consistent views be available for the design representation. Roy and Lui [56] discuss the necessity for varying levels of abstraction when representing tolerances and show methods for maintaining particular low-level entities and determining particular high-level features. The techniques described in this book extend this abstraction capability from predefined hierarchies to any useful relationship described by the user.

Chapter 3

Framework For Tolerance Synthesis

This research proposes an approach to geometric tolerances where function and geometric form are considered simultaneously. The tasks of analysis and synthesis are unified within a common framework that is significantly different from the discrete tasks of analysis and synthesis discussed in the previous chapter. A consequence of this approach is that a significant step is made toward the representation of complete tolerance specifications.

A complete tolerance specification implies that all geometric variations of a product are bounded. Further, the bounds are derived from a set of consistent functional requirements. It is argued here that the tolerances specified for a design describe all product attributes that are negotiated between design and manufacturing concerns. The framework developed in this work provides the representation and metrics that are necessary for concurrent engineering where many different concerns of the product lifecycle are resolved early in the design.

In this chapter, current practices of tolerance specification are first discussed to provide motivation. A framework for tolerance synthesis is then developed and contrasted with the current practice.

3.1 Current Practice

Geometric tolerances are generated by considering the functional requirements for a part and relating these requirements to the geometry of the part under design. The current practice in many design activities is to specify only those tolerances deemed important. Each tolerance is determined via engineering analysis [17] or by using standards [1, 9] that have been developed for common parts and assemblies. All remaining tolerances are then determined according to defaults associated with the drawing or the fabrication process [19]. The specified tolerances along with all of the default tolerances form a tolerance specification that describes all of the variations

CHAPTER 3. FRAMEWORK FOR TOLERANCE SYNTHESIS

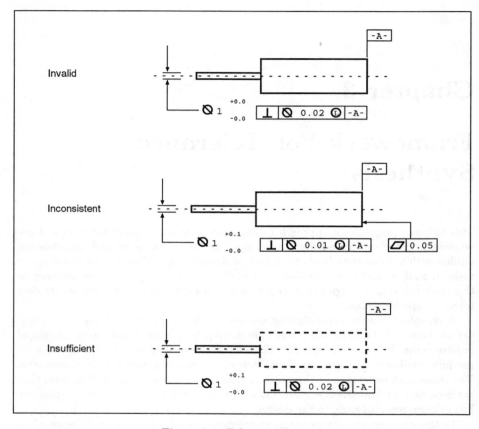

Figure 3.1: Tolerance Errors

that are allowed in the geometry of a part. This approach is expedient but allows several different kinds of errors as illustrated in Figure 3.1.

Unchecked default tolerances may allow variations that contradict the intentions of the designer. As well, tolerances specified by the designer may be unattainable in manufacturing. These are errors of *validity* since the tolerance specification describes parts that do not meet design or manufacturing requirements. The invalid specification of Figure 3.1 calls for ±0.0, or no variation, on a diameter. This tolerance is invalid since it cannot be attained in manufacture.

A precise tolerance may be specified with respect to a geometric feature that is toleranced more loosely. This is an error of *consistency* since the variations allowed by one tolerance do not coincide with the variations allowed by another. At best, consistency errors result in tighter tolerances and higher costs but invalid specifications

3.1. CURRENT PRACTICE

may also result. The inconsistent specification of Figure 3.1 limits a perpendicularity variation to 0.01 while requiring flatness of 0.05 on the corresponding reference datum. These tolerances are inconsistent since the more coarse flatness tolerance influences the finer perpendicularity tolerance.

When a geometric feature is not toleranced at all, manufacturing planning decisions may lead to geometric variations that do not satisfy the functional requirements of the design. This is an error of *sufficiency* since the tolerance specification does not completely communicate design and manufacturing requirements. This error takes on greater importance in computer approaches to tolerance synthesis where redundant checking may not be performed. The insufficient specification of Figure 3.1 includes a cylinder, shown in dashed lines, which is not toleranced at all. The tolerance specification is insufficient since any variation may be allowed for the cylinder.

Previous computer-based approaches to tolerance specification have focused on representation and analysis, i.e. how tolerance information is stored and whether specified tolerances satisfy the functional requirements. Procedures for the synthesis of tolerance specifications require additional knowledge and reasoning. For example, in determining tolerances for a mechanical assembly, one must connect functional requirements with design features, satisfy all relationships between requirements and design features, and ensure that the tolerances prevent undesirable performance. Engineers specifying tolerances may use specialized knowledge and consider only portions of the total design. The tolerance specification produced is considered acceptable if it satisfies functional requirements and falls within manufacturing capabilities. A specification is arrived at by incremental refinement, where particular details are added and then checked, and by global synthesis where many different parameters are chosen together to maximize a particular objective.

In the current practice, tolerancing errors are corrected during design reviews. Prevention is difficult, however, because design and planning documents do not contain information that explicitly connects tolerances with functional requirements and manufacturing capabilities.

Current tolerance analysis techniques can be used to prevent invalid tolerances but the resulting analysis provides little information to resolve conflicts between design and manufacturing. A more detailed tolerance representation is required.

Current tolerance synthesis techniques can be used to choose tolerances that satisfy design and manufacturing requirements but the results generally provide only portions of the tolerance specification and no information about the additional tolerances required to achieve sufficiency while maintaining validity. A procedure for measurement of sufficiency and methods for linking analysis and synthesis are required.

A review of previous research and current practice in tolerance specification suggests that substantial improvements are required. Improvements will only be useful if they address the means by which design is accomplished. Tolerance standards, design standards, manufacturing plans, performance codes, and previous designs, as well as new design and manufacturing features are all used to arrive at new product designs. Methods are required for the composition of complex specifications from a variety of simpler features. Further, while there is great advantage in connecting the many kinds of engineering knowledge employed during a design effort, the many details can be more a burden than a help. Improvements will only be useful if individual engineers can interpret the features that are of interest to them without being expert on the whole design. Methods for abstraction and explanation that accommodate particular engineering disciplines are required.

To address the requirements of tolerance specification, we introduce a tolerance representation based on the VBR theory of tolerances. Using this representation, procedures are developed for the checking of sufficiency and validity. Further, the computation employed demonstrates that tolerance analysis and synthesis can be considered concurrently. Abstraction methods and focused explanation of the representation are developed to manage the many types of knowledge maintained in the representation.

This approach to tolerance specification integrates several different types of engineering knowledge. To effectively use this knowledge integration, we propose a framework for tolerance synthesis.

3.2 A Framework for Tolerance Synthesis

The framework for tolerance synthesis [39], as shown in Figure 3.2, includes tolerance representation, links to functional requirements, validity, sufficiency, explanation, composition, and synthesis. As the figure shows, computational elements, such as sufficiency and validity calculations, and data elements, such as geometric data for nominal solids in the design, are required. Additionally, some data elements (indicated by dashed-lines) depend on the data that is present as well as procedures previously selected for calculation. For example, the data elements in the tolerance representation are only valid when validity and sufficiency conditions are satisfied.

A specific tolerance representation is used in this framework to provide a uniform means of describing how function and form are related. By providing a means for linking functional requirements and geometry, the framework describes gauges that are useful for conformance testing. Apart, these two tasks are interesting for mathematical analysis and computing demonstrations. In combination, they describe software and knowledge that is useful in manufacturing and reusable in design.

To guard against errors of inconsistency, interactive computation is used to check the validity of the tolerance specifications. For efficient determination of design completion, similar computations are provided to test for sufficiency. These tasks could be done separately but computation time and complexity is conserved by combining the effort. The basis for these computations is, of course, the representation and functional requirements of the tolerance synthesis framework.

3.2. A FRAMEWORK FOR TOLERANCE SYNTHESIS

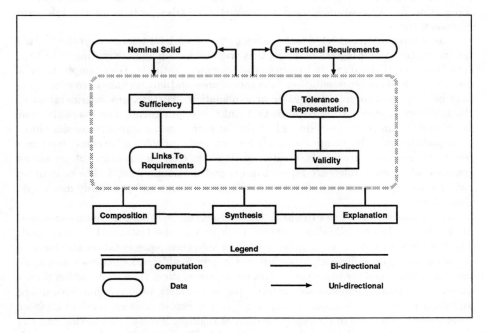

Figure 3.2: A Framework for Tolerance Synthesis

To allow for step by step specification, facilities are provided for composition of tolerance specifications so that more complex descriptions can be built by combining simpler geometric features. At certain stages of the design it is also useful to consider optimizations that account for large parts of the design. Synthesis tools based on optimization techniques provide this capability. While these capabilities could be provided separately, added utility comes from their combination. The composition tools are used to build feasible space for future optimizations and optimization steps provide further features that must be considered during composition. To provide an efficient and understandable connection between these two tasks, feature-based recognition or explanation is used to convert particular portions of the tolerance representation.

These seven interconnected tasks provide a framework for tolerance synthesis. It is the interaction and integration among these representation and computing tasks that contributes most significantly to automatic tolerance synthesis. For example, when a query is made of the representation, tolerance representation and links to requirements may be employed to provide a value or explanation. Further, any consideration of the tolerance representation requires that links to requirements, nominal solid, and functional requirements be evaluated. A similar path must be traced to consider links to requirements. The computing path for synthesis includes sufficiency, tolerance representation, and previously mentioned dependencies. Sufficiency computations require evaluation of tolerance representation and the nominal solid. The computing path for analysis includes validity, tolerance representation and previously mentioned dependencies.

This framework serves as the basis for the tolerance synthesis methods developed in this book. In the following chapters, each part of the framework is developed. The tolerance primitives of Chapter 4 address tolerance representation and links to requirements. The sufficiency procedure of Chapter 5 addresses sufficiency and analysis tasks while making use of the tolerance representation. The composition procedures of Chapter 6 demonstrate methods for maintaining validity while accounting for functional requirements and nominal solids. The explanation approach of Chapter 7 demonstrates queries of the representation at multiple levels of abstraction and the means by which optimizations can be performed. The integration of the framework is further demonstrated in Chapter 8 with several examples.

Chapter 4

Primitives For Reasoning About Tolerances

To simultaneously reason about function and geometric form, we introduce tolerance primitives. A tolerance primitive describes conditional relationships that exist between performance parameters and the geometric form of the design. This approach is based on a rigorously developed theory of tolerance representation.

In this chapter, the tolerance theory which provides a foundation for the tolerance primitives is first discussed. Two distinct classes of tolerance primitives are then described. The first class covers performance requirements that are concerned with geometric relationships such as fit and bulk. Primitives of this class often employ equations that describe geometric and kinematic requirements. The second class covers performance requirements that are characterized by measurements rather than predictive models. Primitives of this class employ tables of measured data.

4.1 Tolerance Theory

A theory for tolerance representation should provide a means for unambiguous and consistent description of tolerances. Further, such a theory should allow for information adequate to test for errors in a tolerance specification. Minimally, a tolerance theory [53] allows for the description of nominal solids, restatement of the nominal solids as features pertinent to tolerancing, and the description of geometric relationships, or tolerance assertions, that must hold in order for tolerances to be satisfied.

Most often, nominal solids such as prisms and cylinders are restated as a collection of surfaces that bound the material-side of the object that is represented. Tolerance assertions are then posed relative to these surfaces or the two half-spaces that each surface defines. Tolerance specifications posed in this way describe a volume, or class of variations, that encloses the boundary of each part that satisfies the tolerances.

Using a tolerance specification based on such a tolerance representation theory, of immediate concern is the question of conformance; *i.e. for a particular specification* and given a manufactured part, is the part in the variational class described by the specification? The tolerance representation must maintain knowledge about geometric shapes and functional requirements as well as gauging criteria to test for conformance.

Distinction is often made between parametric and nonparametric representation for tolerance specifications. The former directly uses the representation of nominal solids to define perfect form tolerance zones. The latter generates tolerance zones by taking the difference between offsets of nominal solids - the solids are used for the difference but their representation does not influence the calculation. Nonparametric approaches are more general and avoid some ambiguity problems. Still, to efficiently measure for conformance, some manner of parameterization is required. It has been suggested by Jayaraman and Srinivasan [27, 62] that a conversion is necessary to practically support measurement of conformance. Using their Virtual Boundary Requirements (VBR) approach, geometric surfaces and relationships between surfaces describe the tolerance specification. To provide for measurement of conformance, the description is then converted to conditional tolerances. The conversion, while applicable to any set of design features, is generally done for commonly occurring features such as primitive geometric solids.

4.1.1 Virtual Boundary Requirements

When specifying tolerances for a mechanical assembly, a designer considers nominal part dimensions, functional requirements, and relationships between the parts of the assembly. The resulting tolerance specification describes geometric constraints for each individual part. More importantly, the geometric constraints for one part in the assembly are independent of the constraints specified for others.

The VBR approach [27, 62] directly supports independence of geometric constraints through the use of virtual surfaces. Virtual surfaces define geometric entities that are shared by features and requirements. For a pair of part features related by an assembly requirement, a virtual surface describes the shared boundary of the pair. For part features related by material bulk requirements, a virtual surface describes bounds on the material of the feature.

For example, Figure 4.1 shows virtual surfaces for a pin and washer assembly. Virtual surface **vs-1** describes the boundary shared by the lip of the pin and the base of the washer. Virtual surface **vs-2** describes the boundary shared by the hole of the washer and the shank of the pin. Figure 4.2 shows virtual surfaces for a tube with material bulk requirements. Virtual surface **vs-3** bounds the inner wall of the tube while virtual surface **vs-4** bounds the outer wall.

Functional requirements are then stated relative to the virtual surfaces. For the pin and washer assembly, assembly is achieved when the shank of the pin is surrounded by the hole of the washer and the lip and base are in close contact. The independent requirements for the pin are that **vs-1** bound the lip, the lip is in close contact with **vs-1**, and that **vs-2** surrounds the shank. The pin can be assembled

4.1. TOLERANCE THEORY

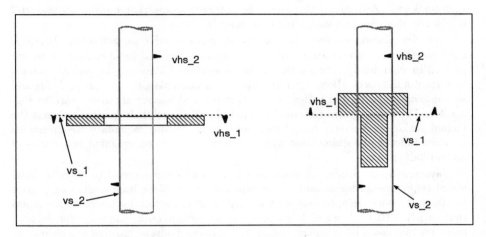

Figure 4.1: Virtual Surfaces for Pin and Washer Assembly

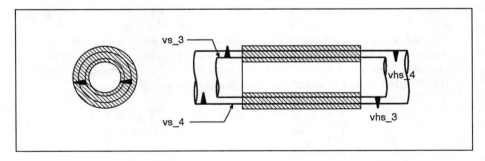

Figure 4.2: Virtual Surfaces for Material Bulk of Tube

when it is possible to position the pin relative to the virtual surfaces and satisfy the independent requirements. Similarly, for the washer, the independent requirements are that **vs-1** bounds the base, the base is in close contact with **vs-1**, and the hole surrounds **vs-2**. As well, for this assembly, there are geometric requirements that **vs-2** is of a specified size and **vs-2** is perpendicular to **vs-1**.

The virtual surfaces, once defined, do not move relative to each other. It is this rigid collection of geometric entities that allows functional requirements to be expressed independently for each part in the assembly. Extending the virtual surfaces to virtual half-spaces allows for formalization of concepts such as *material side* and *surrounding*. The virtual half-spaces for the pin and washer are also shown in Figure 4.1. The arrows point toward the material side of each half-space. Note that the virtual surfaces effectively bound the half-spaces. Virtual Boundary Requirements then are functional requirements stated relative to a rigid collection of boundaries of virtual half-spaces.

Jayaraman and Srinivasan show that these VBRs can be used to describe functional requirements for assembly and material bulk. More important, using offset solids, they have developed necessary and sufficient conditions for tolerance assertions that satisfy VBRs. A set of tolerance assertions made on a nominal solid define a point set that could be specified equally by regularized offsets, growing and shrinking, of the features of the solid. For each offset, one parameter, a_i, describes deviation from nominal. A part instance can be characterized by similar regularized offsets, with parameter b_i, applied to the rigid collection of virtual half-spaces used to describe functional requirements. For assembly, the tolerance assertions are satisfied if and only if, for all i, $b_i \leq a_i$. For material bulk, the tolerance assertions are satisfied if and only if, for all i, $a_i \leq b_i$.

This necessary and sufficient condition allows for a conversion from the VBRs to conditional tolerances which may then be used for measurement of conformance. The assembly requirements of the pin and washer can be considered more generally as assembly for a hole and stud pair. Using parameterized features, as shown in Figure 4.3, the conditional tolerance zone, Z, for assembly of the peg and the conditional tolerance for the orientation angle of the stud, c_3, can be derived [27, 62] as shown in Equations 4.1 and 4.2. Similar results have been derived [27, 62] for position tolerance on the stud/hole pair and orientation and position on key/slot pairs.

$$Z = \left\{ z \;\middle|\; \begin{array}{l} (s_1 > 0) \wedge (l_1 > 0) \wedge (0 \leq c_3 < \frac{\pi}{2}) \wedge \\ (s_1 \sec c_3 + l_1 \tan c_3 - s_{N1} - 2|a_1| \leq 0) \end{array} \right\} \qquad (4.1)$$

$$0 \leq c_3 \leq 2 \arctan \left[\frac{\frac{(s_{N1} + 2|a_1| - s_1)}{l_1}}{1 + \sqrt{1 + \left(\frac{(s_{N1} + 2|a_1| + s_1)}{l_1}\right)\left(\frac{(s_{N1} + 2|a_1| - s_1)}{l_1}\right)}} \right] \qquad (4.2)$$

4.1. TOLERANCE THEORY

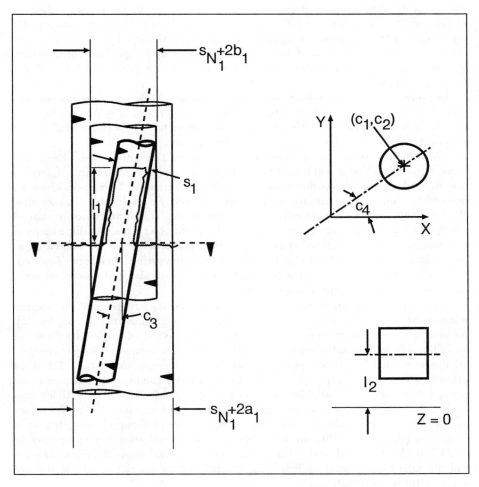

Figure 4.3: Parameterized Stud for Stud/Hole Pair

4.2 Tolerance Primitives

The conditional tolerance relations shown in equations 4.1 and 4.2 are the mathematical equivalent of a functional gauge and can be used directly to measure for conformance. While this, in itself, is a very useful result, the knowledge embedded in the conditional tolerance relations is quite useful *during* the design process. By encapsulating the variables and equations of each conditional tolerance relation, tolerance primitives are created. Tolerance relationships may be specified during design by associating instances of these tolerance primitives with particular features of a nominal solid model.

In general, a tolerance primitive is defined as a set of tolerance assertion parameters $T = \{a_1, ..., a_m\}$, a set of datums $D = \{d_1, ..., d_n\}, 1 \leq n \leq 3$, and, for each member j in the primitive, both a set of fitting parameters $B_j = \{b_{j1}, ..., b_{jo}\}$, and a set of conditional tolerance relations $CT_j = \{ct_{j1}, ..., ct_{jp}\}$.

This representation scheme provides several interesting advantages. Each conversion from functional requirement to conditional tolerance results in a tolerance primitive which describes sufficient conditions for conformance. These primitives can be used by designers as handles (or features) for specifying tolerance relationships. With an appropriate software environment, consistency among tolerance primitives can be evaluated interactively, preventing inconsistent tolerances. As dimensions of the nominal solid are specified or modified, conformance with the conditional tolerance relationships can also be checked interactively, preventing invalid specifications. The relationships specified by the tolerance primitives are also quite useful for measuring sufficiency of tolerance specifications.

This approach differs significantly from other tolerance representation schemes where tolerance ranges and types are stored for particular features [12, 52, 56, 58]. As well, tolerance primitives extend and generalize the system described by Requicha and Chan [55] which used a variational graph embedded in a constructive solid geometry modeler and a facility for specifying and storing tolerance information. Tolerance primitives generate a variational model which includes conditional tolerance equations, datums, tolerance variables, and variables or attributes associated with features of the solid model. By using constraint propagation, tolerance and solid model values can be checked for consistency each time a value is set or changed. Last, tolerance primitives provide flexibility not available with variational geometry approaches to CAD [20]. Dimension-driven CAD systems provide a collection of primitives from which a solid may be built while tolerance primitives are associated with features of a solid after it is defined.

4.2.1 Assembly Requirements and Fit

The tolerance primitives for assembly requirements and fit [85] closely follow the formulations developed by Jayaraman and Srinivasan [27, 62]. The key extension

4.2. TOLERANCE PRIMITIVES

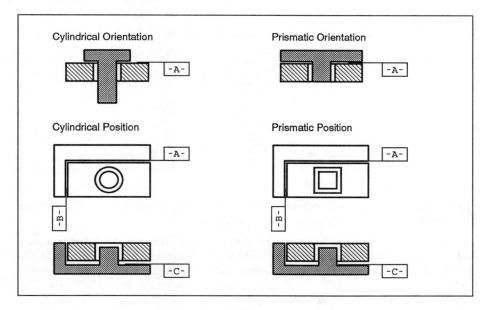

Figure 4.4: Tolerance Primitives for Assembly and Fit

in this work is the formulation and implementation of a computing method for the tolerance primitives.

Tolerance primitives are implemented for both cylindrical and prismatic mating pairs as illustrated in Figure 4.4. The cylindrical orientation primitive describes sufficient and necessary conditions for a cylindrical stud fitting in a cylindrical hole. One datum, labeled **A**, is specified to reference the attitude of each cylinder. The cylindrical position primitive describes sufficient and necessary conditions for a cylindrical stud fitting in a cylindrical hole while positioned according to a complete datum system. Three datums, labeled **A**, **B**, and **C**, are specified to reference the position, attitude, and azimuth of each cylinder. Similarly, the prismatic orientation primitive and prismatic position primitive handle assembly and material bulk for a prismatic slab and prismatic slot.

The prismatic and cylindrical tolerance primitives each have one tolerance assertion parameter, a_1. This parameter describes the allowable variation, via shrinking or growing, for the primitive solid.

One datum is specified for the orientation primitives while three datums are required for the position primitives.

The fitting parameters for each tolerance primitive correspond to a parametric representation used to describe the solid associated with the tolerance primitive. For

Table 4.1: Parametric Space for Cylindrical Solid

s_1	diameter
c_1	x location
c_2	y location
c_3	attitude
c_4	azimuth
l_1	length
l_2	z location

the cylindrical tolerance primitives, seven parameters, as shown in Table 4.1 and Figure 4.3, are used for the parametric representation.

The diameter, s_1, refers to the diameter of a perfect form cylinder fitted to a part instance. On the xy plane, c_1 and c_2 specify the intersection point for the axis of the fitted cylinder. The attitude of the axis, specified by c_3, is the angle between the axis and the z axis. The azimuth, specified by c_4, describes rotation about the z axis. The length of the cylinder is specified by l_1. The location of the midpoint of the fitted cylinder, with respect to the z axis, is specified by l_2. A similar parametric space is employed for prisms.

The conditional tolerance assertions for each tolerance primitive include the equations that describe the conditional tolerance zone for assembly and material bulk. For example, when $s_1 \leq s_{N1} + 2|a_1|$, Equation 4.2 describes the conditional tolerance relationship for assembly of a cylindrical stud in a cylindrical hole when orientation is constrained.

The tolerance primitives are implemented with the constraint definition facilities of the Intelligent Design Environment for Engineering Automation (IDEEA) [24]. Macro constraints are used to encapsulate the most detailed parts of the equations. Functions are used to build constraint networks that check for necessary conditions while enforcing the conditional tolerance relationships.

Table 4.2 shows portions of the definition for the cylindrical orientation primitive. The macro form **tolprim-sol.p** encapsulates the many calculations of Equation 4.2. The **defmath** form specifies that this macro may be referenced in **constrain** forms as **tolprim-sol.m**. The function **assembly-cyl-stud-orient-tol-constraints** applies the orientation tolerance constraints to enforce assembly requirement for a cylindrical stud fitting into a cylindrical hole. The **with-binding-contour** and **constrain-binding-contour** forms manage intermediate variables while constraints are assigned. The attitude parameter, c_3, is assigned no value or a calculated value according to the necessary condition $s_1 \leq s_{N1} + 2|a_1|$. This condition is evaluated by **net?** in the final form of the function.

4.2. TOLERANCE PRIMITIVES

Table 4.2: Example of Cylindrical Fit Primitive

```
(defmac (tolprim-sol.p (a1 c3 l1 s1 sn1)) (a1 c3 l1 s1 sn1)
  (c3 =
    (2.0 * atan( ((sn1 + 2 * abs(a1) - s1) / l1) /
             (1 + sqrt( 1 +
                     ((sn1 + 2 * abs(a1) + s1) / l1) *
                     ((sn1 + 2 * abs(a1) - s1) / l1))))))))

(defmath tolprim-sol.m tolprim-sol.p (sn1 l1 a1 s1) c3 :order 0)

(defun assembly-cyl-stud-orient-tol-constraints (aframe &aux net1 net2 flag)
  (with-binding-contour
      ((frame aframe)
       (no-value (constant nil)))

    (constrain-binding-contour
      ((frame a1 c3 l1 s1 sn1))

    ;; setting each possible route to calculating c3

    (multiple-value-setq (flag net1)
      (constrain tolprim-sol.m( sn1 l1 a1 s1 ) = c3))

    (multiple-value-setq (flag net2)
      (constrain c3 = no-value))

    ;; initial state set to use conditional tol.

    (setf (net-in? net1) t (net-in? net2) nil)

    (with-binding-contour
        ((neta (constant net1)) (netb (constant net2)))

      ;; chose calculation route according to necessary condition

      (constrain sn1 + 2 * abs(a1) = net?(neta netb s1)
              :network "Nec. Cond for tolprim-sol.m"))))))
```

Table 4.3: Possible States for a Necessary Condition

$$s_1 \leq s_{N1} + 2|a_1|$$

State	Example Values		
	s_1	s_{N1}	a_1
true	0.9	1.0	0.01
false	1.1	1.0	0.01
unknown	—	1.0	0.01
unknown	1.1	—	0.01
unknown	1.1	1.0	—
unknown	—	1.0	—
unknown	—	—	0.01
unknown	1.1	—	—
unknown	—	—	—

Using the constraint networks of IDEEA, interactive computations are implemented for the necessary conditions and conditional tolerance relationships of each tolerance primitive. For example, for the orientation of a cylindrical stud fitting in a hole, the necessary condition $s_1 \leq s_{N1} + 2|a_1|$ is evaluated each time s_1, s_{N1}, or a_1 are set or modified. Table 4.3 shows the different states that may be assigned to the necessary condition.

When the state of the necessary condition is true, the equation embedded in tolprim-sol.p is used to enforce the conditional tolerance relationship. When the state of the necessary condition is false, no value may be set for c_3 since a null value is enforced. That is, there is no value for c_3 which is consistent with an assembly requirement for a cylindrical stud fitting in a cylindrical hole. The state of the necessary condition is unknown when values have not been assigned to all parameters. In this case, no conditional tolerance relationships are enforced and other calculations in the constraint network may independently influence the value of c_3.

Since the tolerance primitives are implemented within a network of constraints, other calculations may influence the parameters of a tolerance primitive. For example, a weight restriction might determine the maximum diameter allowed for a cylindrical part. The propagation action of the constraint network allows the values calculated from the weight restriction to serve as input to the tolerance primitive. In this case, the nominal diameter s_{N1} would change or be set. With a new value for s_{N1}, the constraints of the tolerance primitive are activated to test that the new s_{N1} value is consistent with current parameters. If no value was previously associated with s_{N1}, the constraints of the tolerance primitive are activated to calculate values for

4.2. TOLERANCE PRIMITIVES

parameters that are influenced by s_{N1}. For example, if values were associated with s_1, a_1, and l_1, when s_{N1} was set, a value would be calculated for c_3.

This computing behavior allows for the interactive checking of conditional tolerance relationships and necessary conditions each time additional details are added to the design. Further, the constraint propagation model ensures that only related calculations are done.

In some cases, several necessary conditions and conditional tolerance relationships must be managed for a single tolerance primitive. For example, the orientation of a cylindrical hole holding a stud may be determined three different ways as shown in Table 4.4. In some cases, when the diameter is larger than the length, the attitude angle is very loosely bound. This bound is expressed as a closed interval between 0 and $\frac{\pi}{2}$. When the necessary condition is not met, a null value is enforced. Otherwise, **tolprim-s02.p** is used to enforce the conditional tolerance relationship.

By implementing the tolerance primitives with a constraint network, known values can be used to determine unknowns. For many of the tolerance primitives, several different combinations of input variables can be used to calculate an unknown parameter as an output variable. In some cases, however, as illustrated in Equations 4.3 and 4.4, the parameters are confounded and direct calculations are not possible.

$$Z = \left\{ z \;\middle|\; \begin{array}{l} (s_1 > 0) \wedge (l_1 > 0) \wedge (0 \leq c_3 < \frac{\pi}{2}) \wedge \\ (2o + s_1 \sec c_3 + l_1 \tan c_3 - s_{N1} - 2|a_1| \leq 0) \end{array} \right\} \quad (4.3)$$

$$\text{where } o = \sqrt{(c_1 + l_2 \sin c_3 \cos c_4 - c_{N1})^2 + (c_2 + l_2 \sin c_3 \sin c_4 - c_{N2})^2} \quad (4.4)$$

In this case, the conditional tolerance zone for position of a cylindrical stud fitting into a hole is described by the tolerance primitive shown in Table 4.4 and may be used for validity checking only. When the necessary conditions are satisfied, additional calculations are activated to check that parameters of the tolerance primitive are consistent with the conditional tolerance relationships. For this type of equation, the constraint network will never calculate a value for c_3 since the equations require that c_3 be on both sides of the equal sign. Stated in terms of the constraint propagation algorithm, to calculate c_3, c_3 must be known. While it would be possible to employ a numerical relaxation technique to calculate c_3 within this type of tolerance primitive, this contradicts the detailed and complete representation employed for the tolerance synthesis framework. Later in the book, additional methods are developed for tolerance synthesis. These same methods are applicable for cases where the tolerance primitives can't be used for direct calculation.

The tolerance primitives for assembly requirements and fit are complete by definition. That is, each functional requirement implies certain geometric relationships and the presence of these geometric relationships provides sufficient evidence that the functional requirement is satisfied. For example, the requirement for a pair of parts assembling is defined by the variation possible for parts that fit together properly. This variation is specified relative to the geometric properties of primitive solids that

Table 4.4: Example of Several Necessary Conditions

```
(defun assembly—cyl—hole—orient—tol—constraints (aframe &aux net1
                                                        net2 net3 flag)
  (with—binding—contour
   ((frame aframe)
    (no—value (constant nil))
    (upto—halfpi (embedded—value (make—i :low 0.0 :high (/ Pi 2)))))
   (constrain—binding—contour
    ((frame a1 c3 l1 s1 sn1 cond0 cond1 cond2 cond3 cond4))
    (constrain cond0 = s1 <? (sn1 — 2 * abs(a1)))
    (constrain cond1 = s1 >? (sn1 — 2 * abs(a1)))
    (constrain cond2 = s1 <? l1)
    (constrain cond3 = s1 >? l1)
    (constrain cond4 = (sqrt(s1 * s1 — l1 * l1)) <? (sn1 — 2 * abs(a1)))

    (multiple—value—setq (flag net1) (constrain c3 = soft=(no—value)))
    (multiple—value—setq (flag net2) (constrain c3 = soft=(upto—halfpi)))
    (multiple—value—setq (flag net3)
         (constrain tolprim—so2.m( sn1 l1 a1 s1 ) = c3))

    (setf (net—in? net1) nil (net—in? net2) nil (net—in? net3) nil)

    (with—binding—contour
     ((neta (constant net1))
      (netb (constant net2))
      (netc (constant net3)))

     (constrain neta = true—when( cond0 ))
     (constrain netb
                 = true—when(select—and(cond1
                                         select—or (cond2
                                                     select—and
                                                     (cond3 cond4)))))
     (constrain netc = true—when(select—and( cond1 cond2))))))))
```

4.2. TOLERANCE PRIMITIVES

Table 4.5: Example of Confounded Parameters

```
(defmac (tolprim-sp1.p (a1 c1 c2 c3 c4 l1 l2 s1 sn1 cn1 cn2))
  (a1 c1 c2 c3 c4 l1 l2 s1 sn1 cn1 cn2)

  (sn1 =
      2 * sqrt( sqr(c1 + l2 * sin(c3) * cos(c4) - cn1) +
                sqr(c2 + l2 * sin(c3) * sin(c4) - cn2))
      + s1 * sec(c3) + l1 * tan(c3)
      - 2 * abs(a1)))

(defmath tolprim-sp1.m tolprim-sp1.p
  (a1 c1 c2 c3 c4 l1 l2 s1 cn1 cn2) sn1  :order 0)

(defun assembly-cyl-stud-position-tol-constraints (aframe &aux net1 flag)
  (with-binding-contour
   ((frame aframe)
    (no-value (constant nil)))

   (constrain-binding-contour
    ((frame a1 c1 c2 c3 c4 l1 l2 s1 sn1 cn1 cn2
            cond0 cond1 cond2 cond3))
    (constrain cond0 = c3 <? (constant (/ pi 2)))
    (constrain cond1 = c3 >? (constant 0))
    (constrain cond2 = s1 >? (constant 0))
    (constrain cond3 = l1 >? (constant 0))

    (multiple-value-setq (flag net1)
        (constrain tolprim-sp1.m(a1 c1 c2 c3 c4 l1 l2 s1 cn1 cn2 ) = sn1))

    (setf (net-in? net1) nil)

    (with-binding-contour
     ((neta (constant net1))
      (constrain neta =
                  true-when(
                     select-and(
                         select-and(cond0 cond1)
                         select-and(cond2 cond3))))))))
```

define the shape of the design. The assembly of two parts implies some conditions about how part surfaces are arranged. That is, the material of one part will be outside the material of another and in some cases a surface of one part will be in close contact with another. These kinds of conditions are necessary and sufficient. If the parts are assembled, the conditions are satisfied and if the conditions are satisfied, the parts may be assembled.

4.2.2 Empirical Requirements

For many functional requirements it is not always possible to state sufficient conditions only in terms of geometry - that is, geometric conditions that imply by themselves that the functional requirement is satisfied. Bearing designs, for example, are determined not only by geometry but also by operating conditions such as rotation speed and load. If low friction and hydrodynamic lubrication are required functions of the design, a sufficient bearing can only be determined by considering operating conditions along with the geometry of the part.

The tolerance primitive formulation has been extended to accommodate operating conditions such as rotation speed and load. The design parameters used to describe operating conditions must be unambiguous. For example, while a load may refer to force applied at a point or across an area, computation and tradeoffs among several design and manufacturing goals is only accurate if load has a single interpretation. Further, minimum and maximum limits are required for the operating conditions.

Following the tabular format used in the majority of design and manufacturing compilations, tolerance primitives for empirical requirements bound geometric features while satisfying the table entries that correspond to operating conditions. These tolerance primitives are said to represent *empirical* requirements because tabular data, often arrived at empirically, is used in the formulation. Examples from bearing design are used here to demonstrate how operating conditions may be incorporated into tolerance primitives. The formulation, however, applies equally well to situations where manufacturing capability is to be considered.

The H6h6 ISO tolerance table, as shown in Table 4.6, is used to illustrate the representation and computation of these primitives. In the simplest case, manufacturing and design data are handled as a table lookup task. For the H6h6 tolerance table, when a diameter or range of diameters is under consideration, the accompanying tolerance range can be arrived at by taking one table entry or merging several. Conversely, when the tolerance is already bounded, tolerance primitive calculations constrain or generate the feasible range of diameters.

A tabular representation similar to Table 4.6 is maintained for each different tolerance primitive of this type. The approach for selecting and merging values is shown in Figure 4.5. When possible, exact entries from the table are used. Otherwise, a conservative bound is used. If the supplied parameters specify a range or set, the

4.2. TOLERANCE PRIMITIVES

Table 4.6: H6h6 ISO Tolerances

Minimum Diameter	Maximum Diameter	Recommended Tolerance
0.00	0.12	.00025
0.12	0.24	.00030
0.24	0.71	.00040
0.71	1.19	.00050
1.19	1.97	.00060
1.97	3.15	.00070
3.15	4.73	.00090
4.73	7.09	.00100
7.09	12.41	.00120
12.41	15.75	.00140
15.75	19.69	.00160

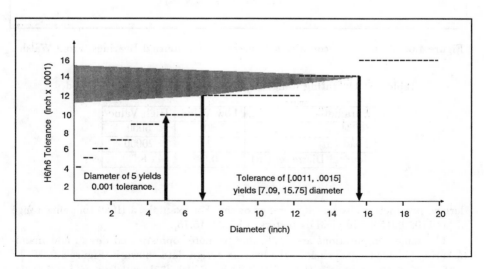

Figure 4.5: Bounding Approximation for Tolerance Primitive

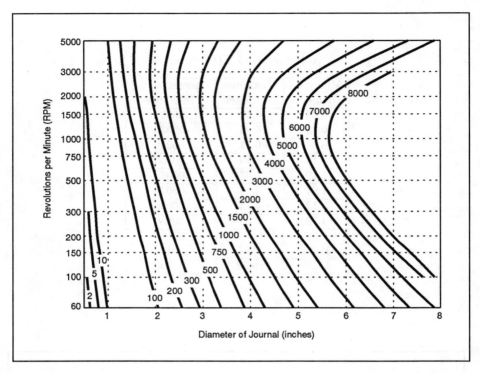

Figure 4.6: Minimum Recommended Diameters for Journal Bearings, From Walsh

Table 4.7: Operating Ranges for Walsh's Bearing Design Method

Parameter	Low Value	High Value
RPM	60	5000
Load (lb)	2	20000
Journal Diameter (in)	0.5	8.0

derived parameters also specify a range or set. For example, a H6h6 tolerance range of [.00110, .00150] will yield diameters of [7.09, 15.75].

The same computations are applicable to more sophisticated design and manufacturing knowledge [86] such as the bearing design data shown in Figure 4.6.

Following the design method described by Walsh [77], satisfactory bearing designs employing hydrodynamic lubrication can be determined for the operating ranges shown in Table 4.7.

4.2. TOLERANCE PRIMITIVES

Table 4.8: Best Engineering Practice for Common Journal Bearings

Width to Diameter Ratio	$b/d = 0.6$
Clearance	$C_d = (0.0009 + \frac{RPM}{5000000})d$
Lubricant	Heavy Machine Oil (30 cP)
Tolerance	H6/h6 or better
Surface Finish	16 microinches

The design method makes use of several broad approximations and some tabular values which have been tuned to reflect experimentally determined behavior. For commonly used journal bearings, best practice design suggests that the ratio of bearing width to journal diameter be close to 0.6. If the lubricating medium is assumed to be similar to heavy machine oil, clearance for the bearing can be approximated as a function of RPM and journal diameter. For these types of designs, surface finish of 16 microinches is sufficient. Walsh suggests that tolerances on the bearing and journal should be of H6/h6 grade or better. Tolerances are assumed to be unilateral with nominal clearance between the journal and bearing as specified by the calculated clearance. These assumptions are summarized in Table 4.8.

The tolerance primitive for bearing design has one tolerance assertion parameter, a_1, which describes shrinking and growing for the primitive solid.

One to three datums may be specified according to the amount of position constraint imposed on the primitive cylindrical solid.

The fitting parameters include the variables used in the parametric representation of the cylindrical solid.

The operating condition parameters include W, and N, the load and rotational speed selected for the design.

The conditional tolerance assertions include those required for assembly of a cylindrical pair and the additional requirements posed on journal and bearing by the Walsh design method.

The nominal journal diameter is constrained by the data of Figure 4.6. The measured length and nominal diameter are also constrained by the width to diameter ratio rule. The tolerance allowed for the journal is determined from the H6/h6 table of the ISO standard fits. The intent is to enforce a unilateral tolerance with respect to the nominal journal diameter. That is, the measured size of the journal may shrink by the H6/h6 tolerance but the measured size of the journal may never exceed the nominal diameter.

These requirements are defined with respect to the parameters of a cylinder as defined for the assembly requirements. For consistency, the unilateral tolerance is first transformed into a bilateral tolerance. For example, the journal diameter d is associated with the nominal diameter of the cylinder s_{N1} and the journal width b

is associated with the measured length l_1 of the cylinder. The tolerance allowed for the journal is associated with the tolerance fitting parameter a_1 of the cylinder. The conditional tolerance assertions for the journal are show in Equations 4.5–4.7.

$$s_{N1} + \frac{1}{2}H6/h6Tolerance(d) \geq WalshbearingDiameter(W, N) \qquad (4.5)$$

$$s_{N1}/l_1 \cong 0.6 \qquad (4.6)$$

$$a_1 \leq \frac{1}{4}H6/h6Tolerance(d) \qquad (4.7)$$

The nominal bearing diameter is also constrained by the nominal journal diameter and the clearance calculated from the journal diameter and RPM. That is, the nominal diameter for the bearing is determined by adding a calculated clearance to the nominal journal diameter. The tolerance allowed for the bearing is intended to enforce a unilateral tolerance with respect to the nominal bearing diameter. That is, the measured size of the bearing may grow by the H6/h6 tolerance but the measured size of the bearing may never be smaller than the nominal bearing diameter. Measured length and tolerance are similar to that of the journal. The conditional tolerance assertions for the bearing are show in Equations 4.8–4.10.

$$s_{N1} - \frac{1}{2}H6/h6Tolerance(d) \geq WalshbearingDiameter(W, N) *$$

$$(1 + (0.0009 + \frac{N}{5000000})) \qquad (4.8)$$

$$s_1/l_1 \cong 0.6 \qquad (4.9)$$

$$a_1 \leq \frac{1}{4}H6/h6Tolerance(d) \qquad (4.10)$$

Note that the bearing cylinder and its accompanying variables are distinctly different from the cylinder defined for the journal. That is, the s_{N1} of the bearing does not equal the s_{N1} of the journal.

The graph of RPM, load and journal diameters, as shown in Figure 4.6, is then used along with some additional calculations to select appropriate parameters for a bearing design. For example, requirements of a 1000 RPM rotational speed and a load of 1000 lb lead to a minimum journal diameter of 2.8 inches. The minimum clearance is then calculated as 0.0038 inches and an H6/h6 is required. For a 2.8 inch diameter, in turn, H6/h6 implies unilateral tolerance of 0.0007 on the shaft or journal and -0.0007 on the hole or bearing.

The tolerance primitive for bearing designs can be used to determine any of the design parameters. When two of the three parameters are specified for the design table, the third entry can be bounded. The algorithm used for this bounding progressively prunes the parameter space. For the three dimensional case, a feasible two dimensional space is first generated by considering one of the specified parameters. A

4.3. CLOSURE

one dimensional space is then generated by considering the second parameter. Finally the feasible space is regularized to yield a real number or closed real interval. This method extends easily to multi-dimensional cases where the performance envelope is convex.

To implement the bearing design tolerance primitive a three column table of load, rpm and diameter, was generated from the graph and 200 table entries were used to approximate the relationships shown in the graph. Increasing the number of table entries would increase the precision of the constraint. Additional calculations for clearance ratio and length to width ratio were added as necessary conditions for the tolerance primitive.

The bearing design tolerance primitive is implemented using the constraint definition facilities of IDEEA. A constraint primitive is defined for the table of bearing design data. Functions are used to select table entries.

Table 4.9 shows the constraint primitive definition for the bearing design tolerance primitive. An additional constraint **walshbearingtable** is defined with three variables. Three calculation rules are defined for the constraint. When **rpm** and **load** are known, diameter, or **diam** is calculated with the form which is a function (**walsh-table-diam rpm load**). The other rules operate in a similar fashion.

Table 4.9 also shows the functions used to query the bearing design table. The functions **get-all-table-values** and **get-first-or-overlapping-table-values** implement the pruning algorithm described earlier.

4.3 Closure

Tolerance primitives have been defined to allow for a common representation of functional requirements and geometric form. These tolerance primitives define the conditional tolerance relationships that must be enforced to ensure that functional requirements are met while preserving appropriate geometric form. The Virtual Boundary Requirements approach to tolerance representation is used as a basis for the tolerance primitives. The IDEEA environment is used to implement each tolerance primitive as a network of constraints. Tolerance primitives are defined for performance requirements that are concerned with geometric relationships and for performance requirements that are determined by operating conditions and empirical data.

In the framework for tolerance synthesis, tolerance primitives provide a means for tolerance representation and for linking functional requirements to part geometry. Chapters 5,6, and 7 also demonstrate that tolerance primitives serve well for measuring sufficiency, insuring validity, and maintaining the detailed knowledge that is needed for analysis and synthesis of tolerance specifications.

Table 4.9: Example of Bearing Design Primitive

```
(defprim (walshbearingtable
          (walshdiam walshrpm walshload)) (diam rpm load)
         (diam (rpm load) (walsh—table—diam rpm load))
         (rpm  (load diam) (walsh—table—rpm load diam))
         (load (rpm diam) (walsh—table—load rpm diam)))
(defun walsh—table—diam (rpm load)
  (return—number—or—interval—from—list
   (get—first—or—overlapping—table—values
    load
    (get—all—table—values rpm *cascade—walsh—table*
                          :index—func #'car
                          :value—func #'pass—thru)
    :index—func #'second
    :value—func #'third)))

(defun walsh—table—rpm (load diam)
  (return—number—or—interval—from—list
   (get—first—or—overlapping—table—values
    diam
    (get—all—table—values load *cascade—walsh—table*
                          :index—func #'second
                          :value—func #'pass—thru)
    :index—func #'third
    :value—func #'first)))

(defun walsh—table—load (rpm diam)
  (return—number—or—interval—from—list
   (get—first—or—overlapping—table—values
    rpm
    (get—all—table—values diam *cascade—walsh—table*
                          :index—func #'third
                          :value—func #'pass—thru)
    :index—func #'first
    :value—func #'second)))
```

Chapter 5
Analysis Tasks and Sufficiency

A sufficient tolerance specification includes a set of tolerance assertions that bound all motion, size, and form of the surfaces of the design. Sufficiency, then, is a measure of the completeness of a design. Sufficient tolerance specifications describe part geometry of unambiguous position and, though ambiguous, the variations in form and size are strictly bounded and assumed to be small.

Determination of tolerance sufficiency is imperative for tolerance analysis where small perturbations are made to, what is assumed to be, a fixed geometrical form. For the development of significant computer-aided tolerance specification tools, the sufficiency metric provides the minimal terminating condition that must be satisfied as a tolerance specification is sought through computation.

In this chapter, tolerance analysis is first discussed as a motivation for sufficiency metrics. Next, a formulation is presented for the sufficiency problem. Several approaches for determination of sufficiency via inspection are then discussed. The particular merits of the tolerance synthesis framework that lend to sufficiency measurement are discussed and an inspection method for sufficiency is described.

5.1 Tolerance Analysis

Tolerance analysis requires a tolerance specification and design requirements as input and determines if the variational class defined by the tolerance specification is within the class required by the design requirements. There are several well established procedures for tolerance analysis. These analyses begin by fixing the nominal geometry that is to be considered. In effect, tolerance sufficiency is postulated. Tolerances are assigned to dimensions that are deemed important. Default tolerances are defined to cover all dimensions which have not been explicitly toleranced. One or more dimensions of interest are chosen as analysis variables. All other dimensions are then perturbed within their tolerance ranges to determine if the resulting effect on the analysis variable satisfies the variation allowed by its tolerance.

48 CHAPTER 5. ANALYSIS TASKS AND SUFFICIENCY

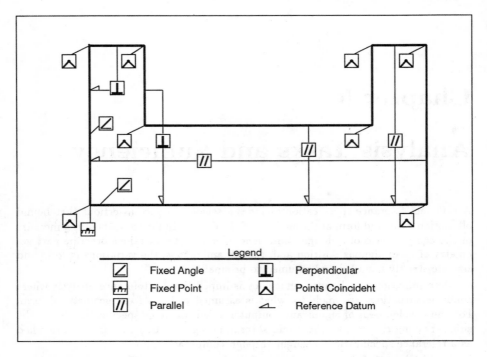

Figure 5.1: Geometric Constraints for Tolerance Analysis Example

Figure 1.3 shows a simple part for tolerance analysis. In fixing the nominal geometry, all points and lines, or features, of the nominal geometry are fully determined. Figure 5.1 shows one means for arriving at a fixed nominal geometry. In this case, the bottom edge of the part is specified as a primary datum and the left most edge is specified as a secondary datum. The point shared by these edges is declared to be a fixed position and each datum is specified at a fixed angle. An angle of ninety degrees separates the two datums. The remainder of the part features are fixed by specifying coincident point, perpendicular, and parallel relationships.

Tolerances are assigned to dimensions and geometric constraints as shown in Table 5.1. In this example, default tolerances are used on all dimensions specified for the part. The collection of geometric constraints illustrated in Figure 5.1 completely constrains the nominal geometry of the part. There are no degrees of freedom among the parameters that specify the points and lines, or features, of the part. Further, there is no redundancy among the constraints.

The Length-1 dimension is chosen as the analysis variable and the corresponding analysis, as presented in Table 1.3 of Chapter 1, suggests that the current tolerance specification is not consistent. Allowable variations in e, c, and d can cause variations

5.1. TOLERANCE ANALYSIS

Table 5.1: Tolerances Assigned in Tolerance Analysis Example

linear dimension	±0.01 inch
fixed angle	0.01 degree
fixed point	0.01 inch in true position
parallel	0.01 inch in zone
perpendicular	0.01 inch in zone

in **Length-1** which exceed the ±0.01 tolerance that is specified. A more detailed discussion of this example, including all of the equations of the tolerance specification, is presented in Appendix B.

The example discussed here employs a linearized sensitivity analysis method for a design described by a consistent and fully determined system of equations. The Mechanical-Advantage® design tool was used for the example. Other techniques [74] are also available for computer-aided tolerance analysis. In a procedural approach, the influences from one model variable to another are defined first and then variations in important parameters are investigated by perturbing dimensions over certain ranges to see how the important parameters are influenced. This is the type of analysis provided in Pro-Engineer®. In a simulation approach, weaker definitions of influences are defined first and then monte carlo simulation is used to generate the actual variations. This is the type of analysis provided in VSA®.

There are several limitations to these tolerance analysis procedures. Functional requirements are not explicitly linked to the geometry of the part. Unless all dimensions related to a functional requirement are checked, there is no guarantee that tolerance analysis by itself will provide a tolerance specification that meets functional requirements and is manufacturable. Analysis must be deferred until all of the geometry can be specified and redundant requirements must be eliminated. The scheme employed for fixing the geometry does not always map directly to achievable manufacturing tolerances or design requirements.

With respect to the tolerance framework that we have described, these tolerance analysis procedures require sufficiency of a tolerance specification in order to evaluate the validity of the specification. The measures used to ensure sufficiency, the independent fixing of all features, do not have any direct correspondence to functional requirements for the part under design. Further, default tolerances are used to ensure sufficiency but are not checked for validity.

A sufficiency measurement is an important component in the framework for tolerance synthesis because it allows design and manufacturing requirements to be the key determiners of how a tolerance specification becomes sufficient. In current approaches to tolerance analysis, the requirements of sufficiency and nonredundancy mean that functional requirements may have to be reformulated or ignored during the analysis.

While tolerance analysis has certainly led to many successful designs, it encourages design errors.

5.2 Formulation for Sufficiency

A sufficient tolerance specification describes bounded variational classes. An unbounded variational class implies several kinds of problems. When an unbounded variational class describes one feature, the position or at least one dimension of the feature is not determined – any value is acceptable. When two or more features are described by an unbounded variational class, there is opportunity to confuse not only size and position but form as well. An unbounded variational class results from an under-constrained specification - from incompleteness or error.

For the specification of tolerance, a criteria for sufficiency is not currently known. It has been suggested by Requicha [53] that sufficiency might be determined by checking that each variational class is strictly bounded and restricted in position.

Following the representation used in the framework for tolerance synthesis described earlier (Chapter 3), sufficiency is defined more directly. All geometric features are described using primitive solids and a parametric representation. All functional requirements are posed using tolerance primitives which directly relate functional requirements and geometric features. Variational classes, bounding the nominal solids, are described by tolerance assertion parameters which define the allowable variation for each parametric solid.

Within the framework for tolerance synthesis, a tolerance specification is sufficient if the specification is valid and there are no degrees of freedom in the system of equations and variables which describe the part geometry and functional requirements of the tolerance specification. A valid tolerance specification, discussed further in Chapter 6, is one in which all conditional tolerance relationships and geometric requirements are consistent. The degrees of freedom in the system of equations and variables are determined as the number of parameters which may be arbitrarily assigned while satisfying all equations of the tolerance specification. Following the definitions and results shown by Sugihara [65] and by Barford [3], this definition can be posed very precisely.

For all of the equations included in a tolerance specification, it is possible to make a set of equivalence classes where the equivalence relationship is that two equations contain exactly the same unknown variables. U refers to the set of equivalence classes and u refers to a particular member of the set. $g(u)$ describes the number of equations that belong to a particular equivalence class. A set of equivalence classes may also be defined for the variables in the tolerance specification where the equivalence relationship is that two variables are contained in the exact same equations. W refers to the set of equivalence classes and w refers to a particular member of the set. $h(w)$ describes the number of variables that belong to a particular equivalence class. For

5.2. FORMULATION FOR SUFFICIENCY

any $Y \subseteq W$, define $\Phi(Y)$ as the subset of U which contains all equations that include any of the variables in Y.

With these definitions, the absence of degrees of freedom is equivalent to the following condition:

$$g(\Phi(Y)) \geq h(Y) \quad \text{for any} \quad Y \subseteq W. \tag{5.1}$$

Inequality (5.1) means that, for any equivalence class of variables in the tolerance specification, the number of equations influencing the variables is greater than the number of variables. In practice, an equivalence class of variables will be the variables which are included in one equation. This definition is a special case of structural rigidity as defined by Barford [3]. For systems of linearly independent equations, this definition is exact since each equation removes one degree of freedom. For systems of nonlinear equations, this is a conservative definition since combinations of equations may remove more degrees of freedom than the number of equations in the combination.

To illustrate how this definition ensures sufficiency, it is useful to consider all of the variables and equations of the tolerance specification. The nominal part geometry is defined as a collection of primitive solids. Each primitive solid has a parametric representation and several equations to define how the parameters of the solid are related. The tolerance primitives use a similar parametric representation for primitive solids and add conditional tolerance relationships which bound the parameter space according to particular functional requirements. Tolerance assertion variables define variational classes for each solid associated with a tolerance primitive by describing the shrinking or growing that may occur for each nominal solid. Functional requirements are defined as equations which relate the parameter space of the nominal primitive solids, the tolerance assertion variables, and the operating condition parameters. Any of the variables in the tolerance specification may also be set according to user preference or standard practice.

Each nominal solid is bounded if its parameter space is bounded – this is guaranteed by constructive solid geometry and boundary representation schemes for solid representation. The variational class associated with each tolerance primitive solid is bounded if its tolerance assertion variables are bounded. In other tolerance representation schemes, the position of each variational class would also be of concern. In the VBR approach all variation is defined relative to a fixed set of coordinates which are related to the nominal geometry. The tolerance assertion variables will be bounded if they can each be determined from at least one set of equations that describes functional requirements.

The set of equations and the set of variables which must be evaluated to measure tolerance sufficiency will be quite large for any significant design. Sugihara has shown [65] that, for many classes of problems, a network flow formulation can be used to solve this problem in polynomial time. Further refinements [3] are possible for linear systems.

This definition is consistent with the definition given by Requicha and allows for tolerance specifications where manufacturing or design requirements further constrain the variational classes defined with respect to primitive solids.

5.3 Inspection Procedure for Detection

An inspection procedure for sufficiency requires, first, a method for determining whether or not tolerance specification variables are bounded. The definition of Equation 5.1 provides one approach for determining boundedness. A particular variable Var_0 can be determined to be bounded by the following algorithm.

Algorithm 5.1 (Boundedness Determination)
    Step 1. If Var_0 is assigned a bound value, return 'bounded'.
    Step 2. Let $Eqns$ be the equations that influence Var_0.
    Step 3. Let $Vars$ be the variables included in the equations of $Eqns$.
    Step 4. Let $Unbound$ be the variables in $Vars$ which have no value.
    Step 5. If the size of $Eqns$ is greater than the size of $Unbound$, return 'bounded'.
    Step 6. Let $SufVars$ be the variables in $Unbound$, excluding Var_0, which are bounded.
    Step 7. If the size of $Eqns$ is greater than the size of $Unbound - SufVars$, return 'bounded'.
    Step 8. Return 'unbounded'.

At most, this algorithm will depend on the number of variables and require $O(n)$ time for the case where all variables are related. Note that during recursion, previous Var_0 instances are not evaluated again as this would cause infinite computation. Recall that this test, based on Equation 5.1, is conservative. It may be useful in some cases to search for algebraic solutions or to converge upon bounds using minimization techniques.

The sufficiency measure suggested by Requicha [53] requires that each variational class be strictly bounded and restricted in position. Using this definition, the sufficiency of a tolerance specification is measured by inspecting each geometric solid in the specification. For each solid, it is first necessary to ensure that position variables are fixed. This may be as simple as a table lookup for each position parameter associated with a geometric solid. However, for solids positioned according to reference datums, it would be necessary to determine which faces of the geometric solid were associated with reference datums. This evaluation is proportional to the product of faces and solids. For collections of simple parts this time would be closer to $O(cn)$ while for complex parts the time required could be closer to $O(n^2)$. For each solid it is also necessary to measure the boundedness of the variables that describe the variational class. If the solid representation and the tolerance specification are

5.3. INSPECTION PROCEDURE FOR DETECTION

maintained separately, the initial setup for each solid requires an evaluation of the entire tolerance specification to find equations which influence the solid. This step is proportional to the product of solids and equations and requires $O(n^2)$ time. The boundedness determination then requires up to $O(n)$ time for each variable.

This inspection procedure is summarized in Algorithm 5.2. The algorithm requires up to $O(n^3)$ time and is proportional to the product of solids and faces plus the product of solids, equations, and variables. Note that this algorithm measures sufficiency only – not validity.

Algorithm 5.2 (Sufficiency Determination I)
Step 1. For each solid in tolerance specification, if position is not fixed, return 'insufficient'.
Step 2. For each variable defined for each solid in tolerance specification, if variable is unbounded (Algorithm 5.1), return 'insufficient'.
Step 3. Return 'sufficient'.

In the framework for tolerance synthesis, consistency is determined incrementally as functional requirements and part geometry are defined. Each time a variable is set to a value, all relevant equations are evaluated to ensure that other variables associated with the equation have values that are consistent. This is accomplished via constraint propagation and may require $O(n^3)$ time since the propagation steps are proportional to the product of variables and the square of equations.

In the framework for tolerance synthesis, all variables for the tolerance specification may be queried and all relationships between variables and equations are represented explicitly. Determining the equations which influence a variable, or vice-versa, is similar to a table lookup and can often be accomplished in nearly constant time. At most, however, this procedure is proportional to the sum of equations and variables and requires $O(n)$ time. For example, if a variable is influenced by 7 equations in a tolerance specification where there are 100 variables and 200 equations, it is likely that two lookup procedure calls will yield a list of 7 pointers which should be evaluated to provide the list of equations. The time required is almost constant when the number of equations is small. In the worst-case, however, all variables could be equated and determined by each equation. In this case all variables and equations would be evaluated to select a set of relevant equations. While this worst-case behavior is possible, it is not relevant to most design problems.

A geometric solid and its variational classes are bounded if the variables of all corresponding tolerance primitives are bounded. Each solid may be evaluated individually by considering only the tolerance primitive variables *Vars* which are associated with the particular geometric solid. This inspection procedure is summarized in Algorithm 5.3. The algorithm requires up to $O(n^3)$ time since the calculation is proportional to the product of variables and the square of equations plus the number of equations plus the number of variables. As well, with the tolerance representation of the framework for tolerance synthesis, each variable can be marked during the evaluation of boundedness, to reduce the time for boundedness determination to $O(n)$. Note that this algorithm measures validity and sufficiency.

Algorithm 5.3 (Sufficiency Determination II)
  Step 1. If the equations which include the variables in *Vars* are inconsistent, return 'insufficient'.
  Step 2. For each Var_i in *Vars*, if Var_i is unbounded, return 'insufficient'.
  Step 3. Return 'sufficient'.

The entire tolerance specification can be considered at one time by evaluating all of the tolerance primitive variables associated with the tolerance specification. Collecting these different sets of variables can be accomplished in $O(n)$ time since the calculation is proportional to the number of solids.

5.4 Closure

Tolerance analysis techniques have been presented to show how sufficiency is currently used *and required* when evaluating tolerance specifications. A priori requirements for tolerance sufficiency make it difficult to link functional requirements with geometric form since most engineering designs include redundant requirements. The emphasis, in the current practice, of sufficiency before analysis also provides the opportunity for many design errors. A sufficiency measurement is important to the framework for tolerance synthesis because it allows the requirements of a design to incrementally constrain a tolerance specification and, ultimately, to determine a sufficient specification.

A definition for the degrees of freedom in a tolerance specification is proposed and an algorithm for determining boundedness is presented. These results are then used to define how sufficiency can be determined via an inspection procedure. An inspection procedure is first described for the approach defined by Requicha [53]. The features of the framework for tolerance synthesis that contribute to sufficiency measurement are then discussed and a second inspection procedure is presented. Both inspection procedures incur significant running time for tolerance specifications that are sufficient. The procedure defined for the framework for tolerance synthesis is superior in that it provides a measure of validity and sufficiency while reducing the computation required to determine fixed positions.

Chapter 6
Synthesis Tasks and Composition

Engineering designs balance many different and, often, competing requirements. When the requirements posed for a design admit at least one design that satisfies the requirements, the requirements are said to be valid. That is, the requirements do not contradict each other.

In our framework for tolerance synthesis, we discussed design methods that allow tolerance requirements to be added to a design incrementally. Here we describe composition techniques where incremental changes in the specification always yield a valid design space [38]. Two distinct cases must be handled for these techniques. Independent requirements are handled with constraint propagation methods. Requirements that are coupled are handled with detection and replacement methods.

In this chapter, design synthesis and composition of tolerance requirements are first described. Independent and coupled cases for composition are discussed and an approach for composition is then presented. Last, examples for the composition method are presented with discussion.

6.1 Composition and Competing Requirements

Design synthesis entails the composition of smaller elements into a whole while satisfying a specification. Performance goals and consistency among elements are of most importance. It is common, when approaching an engineering design, to first select a schematic design using experience and functional requirements and then to develop a mathematical model for the dominant physical properties of the schematic. The schematic and mathematical model are then improved and refined via design synthesis to yield a complete design.

In most cases, the mathematical model describes output or performance that can be expected from particular inputs or controls. For example, when material A is selected and geometry is described by C, D, and E, stiffness is predicted to be F. In synthesizing a design, however, the goal is often to find appropriate parameters, such as material and geometry, so that an output, such as stiffness, remains at some specified value. In effect, the synthesis task requires that the mathematical model be inverted.

Tolerance synthesis involves the specification of tolerances to ensure design performance, reliability, and efficient fabrication. In the course of a design, a variety of tolerance requirements are associated with the part under design. In practice, it is common to add dominant requirements early and then check the feasibility of these requirements as more detailed items are added.

The tolerance primitives defined in Chapter 4 describe and enforce relationships that exist between functional requirements or design performance and tolerances. These tolerance primitives are instanced in a constraint network and associated with the nominal geometry for a design. Each time a tolerance primitive is added for a particular functional requirement, additional tolerance assertions are associated with the design. The tolerance primitives allow for the incremental development of a tolerance specification directly from the functional requirements.

A valid tolerance specification describes variational classes for parts which satisfy design and manufacturing requirements. The first step in tolerance synthesis is to arrive at a valid tolerance specification but, in practice, the tolerance assertions required for all functional requirements in a design may be contradictory. That is, one functional requirement may demand a larger variational class for a geometric feature that was previously bounded by a smaller variational class determined from another functional requirement.

A significant part of the tolerance synthesis problem is the determination of a valid tolerance specification. Tolerance primitives allow this to be determined through a series of incremental compositions. Each time an additional functional requirement is described with tolerance primitives, a new composition is formed. The composition is valid if all functional requirements related to the tolerance primitives are simultaneously satisfied.

6.2 Independent and Coupled Functional Requirements

When composing tolerance primitives to form a tolerance specification, a valid tolerance specification is maintained as long as all of the conditional tolerance relationships of the tolerance primitives are consistent. To maintain validity of the tolerance specification, two different cases of composition must be handled.

In the first case, each functional requirement is independent. That is, the conditional tolerance relationships that are defined for the tolerance primitive influence only the tolerance assertion parameters and fitting parameters associated with the particular tolerance primitive.

Figure 6.1 illustrates a simple example of independent tolerance primitives. The assemblies labeled A and B correspond to cylindrical position tolerance primitives. If

6.2. INDEPENDENT AND COUPLED FUNCTIONAL REQUIREMENTS

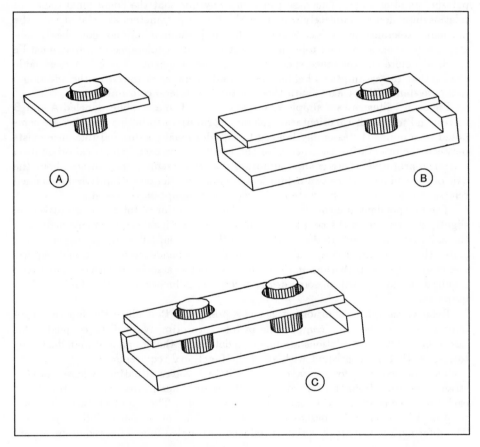

Figure 6.1: Composition of Independent Primitives

both assemblies are composed to produce the assembly labeled C, the earlier tolerance requirements remain the same. Each cylindrical stud and hole is positioned relative to a datum system defined by the lower part.

In the second case, functional requirements are coupled and parameters of one tolerance primitive may be influenced by the conditional tolerance relationships and parameters of another. This case can result anytime that the conditional tolerance relationships don't completely constrain the fitting parameters associated with the particular tolerance primitive. When the fitting parameters are not completely constrained by their associated tolerance primitive, another tolerance primitive must be specified before the tolerance specification can be sufficient. There is a reasonable chance that the assumptions behind the second tolerance primitive or the additional datum system will not agree with those of the first tolerance primitive.

Figure 6.2 illustrates a simple example of this. The assemblies labeled A and B correspond to cylindrical orientation tolerance primitives. If both assemblies are composed to produce the assembly labeled C, an additional conditional tolerance exists between the orientation parameters of the two fits. For each cylindrical orientation tolerance primitive, there is an assumption that the assembly may rotate about the axis of the cylinder. Once two cylindrical orientation tolerance primitives are defined relative to the same datum system, the rotation assumption is invalid.

The independent and coupled cases for the composition of tolerance primitives are slightly different from the coupled and uncoupled functional requirements defined in the axiomatic approach to design developed by Suh [66]. The independence axiom states that an optimal design always maintains independence of functional requirements such that perturbations in the design parameters associated with one functional requirement do not, within some design tolerance, influence any other functional requirements.

Relative to Suh's definitions, there is a possibility that all of the functional requirements described in a composition of tolerance primitives will be coupled. This has to be so when calculations are made to determine the design and manufacturing tolerances that are consistent with all of the functional requirements.

The coupled case for tolerance composition involves situations where the constraints resulting from two tolerance primitives are different from the constraints for each tolerance primitive when considered individually. The approach taken for composition in this case, is consistent with the information axiom of Suh's approach. The information content of design is reduced as coupled functional requirements are considered together.

6.3 Approach for Composition

Independent functional requirements are easily composed in the framework for tolerance synthesis by using constraint networks and propagation. Constraint-based reasoning is appealing for engineering tasks because it closely matches problem-solving

6.3. APPROACH FOR COMPOSITION

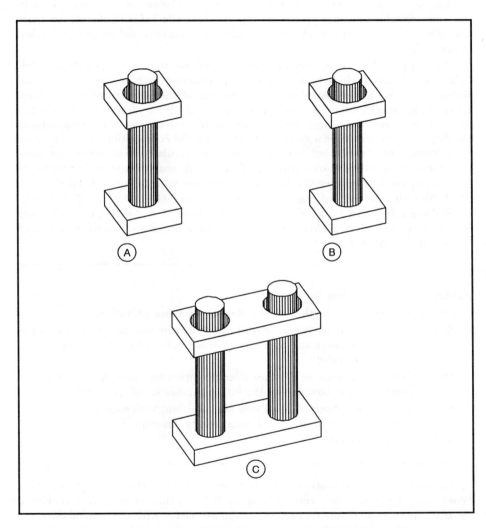

Figure 6.2: Composition of Primitives with Coupling

strategies employed by practicing engineers. Engineers often design by first determining unknowns from dominant relations and known quantities. Later in the design process, details are added and checked for validity. The constraint system used in the framework for tolerance synthesis supports these two levels of abstraction quite naturally while providing mechanisms to preserve, identify, and explain numerical relationships.

Composition for independent functional requirements requires two stages. Initially, the tolerance primitive is specified and associated with one or more nominal geometric solids. In the second stage, the validity of the composition is evaluated. As values are specified for the tolerance assertion and fitting parameters of the tolerance primitive, constraint propagation is used to test the validity of the composition. Validity is tested each time a value is specified. Invalid compositions may remain in the tolerance specification but parameters are only calculated from parameters associated with valid compositions. Also, invalid compositions are constantly brought to the attention of the designer via the contradiction handling features of IDEEA.

Composition for coupled functional requirements requires four stages. First, the coupled condition must be detected. This can be accomplished within the framework for tolerance synthesis by making two tests each time a tolerance primitive is instanced as show in Algorithm 6.1.

Algorithm 6.1 (Coupling Detection)

Step 1. Let *NewPrim* be the newly instanced tolerance primitive.

Step 2. If the conditional tolerance relations *NewPrim* completely constrain the fitting parameters and tolerance assertion parameters of *NewPrim*, return 'uncoupled'.

Step 3. Let *OtherPrims* be all other tolerance primitives associated with the nominal solids bounded by the fitting parameters of *NewPrim*.

Step 4. If the assumptions of *NewPrim* are consistent with each tolerance primitive included in *OtherPrims*, return 'uncoupled'.

Step 5. return 'coupled'.

Coupled tolerance primitives must then be disabled in the constraint network. When using the constraint representation of IDEEA, this amounts to a change of the truth-maintenance state for each constraint associated with the coupled tolerance primitives. The coupled tolerance primitives must then be replaced with a different tolerance primitive that enforces the functional requirements intended by the designer and accurately describes the coupling condition. Once the coupling is resolved, validity of the composition is then evaluated as described for independent functional requirements.

6.4. EXAMPLE FOR INDEPENDENT FUNCTIONAL REQUIREMENTS

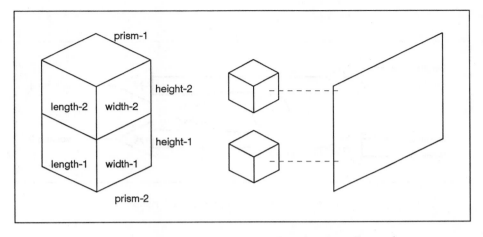

Figure 6.3: Independent Functional Requirements Example

6.4 Example for Independent Functional Requirements

To illustrate how composition is handled for independent functional requirements, it is useful to consider a very simple example of relations between prisms. As shown in Figure 6.3, **prism-1** and **prism-2** each have parameters **height**, **width**, and **length**. As well, relations are known to describe **area** and **volume** for each prism. Assume that the prisms share a common face and that the width of both prisms must be the same.

Relative to the tolerancing problem, the prisms correspond to features of a part under design or tolerancing primitives. The **area** and **volume** relations are validity measures that must be enforced at all times for the feature or primitive. The shared face and width are relations that ensure that a functional requirement is satisfied.

The prism features can be represented in a network of constraints as shown in Figure 6.4. For each prism, **area** is calculated from **length** and **height** while **volume** is calculated from **area** and **width**. In the constraint network, the calculations can occur in any direction. That is, **length** can be determined if **area** and **height** are known.

As values are associated with each parameter, constraint propagation is used to evaluate the validity of the specification. For example, Table 6.1 shows starting values for some of the parameters in the constraint network of Figure 6.4.

These starting values determine values for the remaining parameters as shown in Table 6.2. Table 6.3 shows a trace of the constraint propagation rules that were used to arrive at these values. When a rule does not yield a new value for assignment, DISMISS is returned and a propagation path is terminated. When a rule does yield a new value, the value is assigned to the parameter and the parameter may activate

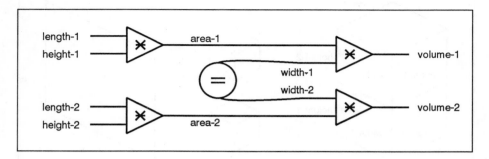

Figure 6.4: Constraint Network for Prisms

Table 6.1: Starting Values for Prisms

length-1	[1,2]
height-1	[1,4]
length-2	[1,3]
volume-1	[20,50]
volume-2	[20,50]

Table 6.2: Determined Values for Prisms

height-2	[0.133,20]
area-1	[1,8]
area-2	[0.4,20]
width-1	[2.5,50]
width-2	[2.5,50]

6.4. EXAMPLE FOR INDEPENDENT FUNCTIONAL REQUIREMENTS 63

additional rules which open new propagation paths. Propagation ceases when all paths have been exhausted.

The DISMISS result can occur for several different reasons. Table 6.4 shows the rule definitions for a multiplication constraint. Several of the rules are used to assign values when one of the parameters has a zero value. For example, when a nonzero value exists for m1, the zero rule will fail and return DISMISS. A rule will also fail if the values determining a new value for a parameter depend on the previous value of the parameter currently being calculated. Finally, a rule will fail if a very small change is calculated for a parameter. These actions prevent infinite calculation when there are cycles in the constraint network. Note that in Table 6.3, some rules have been removed from the trace. The removed rules showed failed attempts to compute a value when one input was zero.

The trace shown in Table 6.3 demonstrates the constraint propagation action. As values are available to supply a computation, new values are calculated. For example, when height-1 is assigned the interval [1,4], as shown in line 3, the first rule evaluated attempts to calculate a value based on M2 being zero. Since height-1 is not zero, this rule fails. The next rule evaluated calculates a product from the values associated with height-1 and length-1. This value is then assigned to area-1. Since area-1 is now assigned a value, several additional rules are activated for the constraint relating area-1, width, and volume-1. Two of these rules fail in an attempt to calculate a value based on M2 being zero. The third fails because no value is assigned to the other required parameters.

As other requirements are added to this specification and as the parameters are modified, the specification remains valid as long as the constraints of Figure 6.4 are satisfied. For example, if an interval of [1,3] is later specified for width-1, a contradiction will be signaled because the new interval for width-1 is not consistent with the previously calculated value of [2.5,50].

More precisely, width-1 was previously determined from the values for volume-1, height-1, and length-1. The intersection of these two values, [1,3] and [2.5,50], is [2.5,3.0] which is the value finally assigned to width-1. This new value for width-1 is not entirely consistent with the values assigned for volume-1, height-1, and length-1. As the interval for width-1 decreases, to satisfy the equations of the constraint network, at least one of the volume-1, height-1, and length-1 intervals must be reduced. In this example, volume-1 was selected for retraction and a new smaller interval was calculated, as shown in table 6.5, for volume-1 via constraint propagation as shown in table 6.6.

The result of this constraint propagation is a feasible parameter space. Most classical optimization techniques start from a feasible parameter space or include a step to ensure that the search space is feasible. The application of constraint propagation to the composition and synthesis of tolerance specifications provides several advantages beyond the benefits of classical optimization.

Table 6.3: Trace From Initial Prism Calculations

```
;; length-1 = [1,2] (3 rules yielding (@DISMISS) omitted)
;; height-1 = [1,4]
;|Running rule <PRODUCT<=*(M2)> on <#:*8871:MULTIPLIER>.
;|Rule <PRODUCT<=*(M2)> returned (@DISMISS).
;|Running rule <PRODUCT<=*(M1,M2)> on <#:*8871:MULTIPLIER>.
;|Rule <PRODUCT<=*(M1,M2)> returned [1.0,8.0].
;|Running rule <M1<=/(M2,PRODUCT)> on <#:*8871:MULTIPLIER>.
;|Rule <M1<=/(M2,PRODUCT)> returned (@DISMISS).
;|Running rule <PRODUCT<=*(M2)> on <#:*8875:MULTIPLIER>.
;|Rule <PRODUCT<=*(M2)> returned (@DISMISS).
;|Running rule <PRODUCT<=*(M1,M2)> on <#:*8875:MULTIPLIER>.
;|Rule <PRODUCT<=*(M1,M2)> returned (@DISMISS).
;|Running rule <M1<=/(M2,PRODUCT)> on <#:*8875:MULTIPLIER>.
;|Rule <M1<=/(M2,PRODUCT)> returned (@DISMISS).
;; length-2 = [1,3] (3 rules yielding (@DISMISS) omitted)
;; volume-1 = [20,50]
;|Running rule <M2<=/(M1,PRODUCT)> on <#:*8875:MULTIPLIER>.
;|Rule <M2<=/(M1,PRODUCT)> returned (@DISMISS).
;|Running rule <M1<=/(M2,PRODUCT)> on <#:*8875:MULTIPLIER>.
;|Rule <M1<=/(M2,PRODUCT)> returned [2.5,50.0].
;|Running rule <RIGHT<==(LEFT)> on <=:=>.
;|Rule <RIGHT<==(LEFT)> returned [2.5,50.0].
;|Running rule <PRODUCT<=*(M1)> on <#:*8877:MULTIPLIER>.
;|Rule <PRODUCT<=*(M1)> returned (@DISMISS).
;|Running rule <PRODUCT<=*(M1,M2)> on <#:*8877:MULTIPLIER>.
;|Rule <PRODUCT<=*(M1,M2)> returned (@DISMISS).
;|Running rule <M2<=/(M1,PRODUCT)> on <#:*8877:MULTIPLIER>.
;|Rule <M2<=/(M1,PRODUCT)> returned (@DISMISS).
;; volume-2 = [20,50]
;|Running rule <M2<=/(M1,PRODUCT)> on <#:*8877:MULTIPLIER>.
;|Rule <M2<=/(M1,PRODUCT)> returned [0.4,20.0].
;|Running rule <M1<=/(M2,PRODUCT)> on <#:*8877:MULTIPLIER>.
;|Rule <M1<=/(M2,PRODUCT)> returned (@DISMISS).
;|Running rule <M2<=/(M1,PRODUCT)> on <#:*8873:MULTIPLIER>.
;|Rule <M2<=/(M1,PRODUCT)> returned [0.1333,20.0].
;|Running rule <M1<=/(M2,PRODUCT)> on <#:*8873:MULTIPLIER>.
;|Rule <M1<=/(M2,PRODUCT)> returned (@DISMISS).
```

6.4. EXAMPLE FOR INDEPENDENT FUNCTIONAL REQUIREMENTS

Table 6.4: Multiplication Constraint Definition

```
(defprim (multiplier (* / /)) (product m1 m2)
  (product (m1) (if (csend m1 zerop) (make—num :num 0) @dismiss))
  (product (m2) (if (csend m2 zerop) (make—num :num 0) @dismiss))
  (product (m1 m2) (csend m1 * m2))
  (m2 (m1 product) (if (not (csend m1 zerop))
                          (csend product / m1)
                          @dismiss))
  (m1 (m2 product) (if (not (csend m2 zerop))
                          (csend product / m2)
                          @dismiss)) )
```

Table 6.5: Modified Values for Prisms

length-1	[1,2]
height-1	[1,4]
length-2	[1,3]
volume-1	[2.5,3.0]
volume-2	[20,50]
height-2	[2.22,20]
area-1	[1,8]
area-2	[6.67,20]
width-1	[2.5,3]
width-2	[2.5,3]

Table 6.6: Trace From Modification of Prism Parameters

```
;; width−1 = [1,3] and retract volume−1
;|Running rule <PRODUCT<=*(M1)> on <#:*8875:MULTIPLIER>.
;|Rule <PRODUCT<=*(M1)> returned (@DISMISS).
;|Running rule <PRODUCT<=*(M1,M2)> on <#:*8875:MULTIPLIER>.
;|Rule <PRODUCT<=*(M1,M2)> returned [2.5,24.0].
;|Running rule <M2<=/(M1,PRODUCT)> on <#:*8875:MULTIPLIER>.
;|Rule <M2<=/(M1,PRODUCT)> returned (@DISMISS).
;|Running rule <RIGHT<==(LEFT)> on <=:=>.
;|Rule <RIGHT<==(LEFT)> returned [2.5,3.0].
;|Running rule <PRODUCT<=*(M1)> on <#:*8877:MULTIPLIER>.
;|Rule <PRODUCT<=*(M1)> returned (@DISMISS).
;|Running rule <PRODUCT<=*(M1,M2)> on <#:*8877:MULTIPLIER>.
;|Rule <PRODUCT<=*(M1,M2)> returned (@DISMISS).
;|Running rule <M2<=/(M1,PRODUCT)> on <#:*8877:MULTIPLIER>.
;|Rule <M2<=/(M1,PRODUCT)> returned [6.6667,20.0].
;|Running rule <M2<=/(M1,PRODUCT)> on <#:*8873:MULTIPLIER>.
;|Rule <M2<=/(M1,PRODUCT)> returned [2.2222,20.0].
;|Running rule <M1<=/(M2,PRODUCT)> on <#:*8873:MULTIPLIER>.
;|Rule <M1<=/(M2,PRODUCT)> returned (@DISMISS).
```

A classical optimization approach for tolerance specification would require that all requirements be stated before a search for the final specification. By using the framework for tolerance synthesis and constraint propagation, the tolerance specification can be incrementally built from functional requirements. As each requirement is added, the validity of the specification is checked. Further, in a classical optimization approach, it is often difficult to determine which modifications are required to move from an infeasible specification to a feasible or valid specification. Using the framework for tolerance synthesis and constraint propagation, useful information is available at each design step to guide design changes when a specification becomes invalid.

The trace of constraint propagation rules in Tables 6.3 and 6.6 suggest that in some cases each parameter associated with a constraint may activate calculations for all other parameters in the constraint network. For example, if all parameters were instantaneously assigned values which were of equal priority in the constraint network, constraint propagation may require up to $O(n^3)$ time. The failure criteria used for constraint propagation, however, prune many of the propagation paths. More so, while engineering designs involve many interrelated details, most constraint networks which describe design specifications are quite sparse.

6.5 Example for Coupled Functional Requirements

Three design steps for a simple part, as shown in Figure 6.5, are used to illustrate how coupled functional requirements can be handled in the framework for tolerance synthesis. During a design session, an engineer associates the cylindrical orientation tolerance primitive with a feature of a solid model such as the stud atop a prism shown in detail A of Figure 6.5. The conditional tolerance relationships for assembly described by Equation 6.1 are then enforced using a constraint network. Note that, for the parameterized tolerance primitive, only 3 of its 7 parameters are constrained.

$$0 \leq c_3 \leq 2 \arctan \left[\frac{\frac{(s_{N1}+2|a_1|-s_1)}{l_1}}{1 + \sqrt{1 + \left(\frac{(s_{N1}+2|a_1|+s_1)}{l_1}\right)\left(\frac{(s_{N1}+2|a_1|-s_1)}{l_1}\right)}} \right] \quad (6.1)$$

Adding a cylindrical orientation tolerance primitive, the design might lead to a centerline distance between the two primitives as shown in detail B of Figure 6.5. The conditional tolerance zones for the two primitives are then coupled. This coupling could be detected by determining that the assumed rotation about the cylindrical axis of each stud is no longer feasible. The constraints enforcing the conditional tolerance relationship of Equation 6.1 must be disabled.

A conservative characterization of the two zones, for assembly, is expressed with Equations 6.2 – 6.6. Parameters c_1, c_2, etc., refer to stud instance one while param-

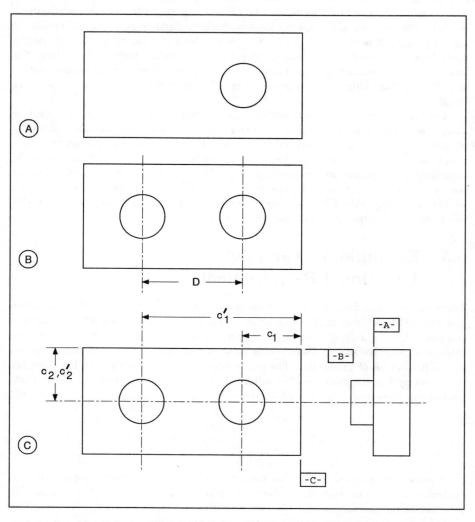

Figure 6.5: Coupled Functional Requirements Example

6.5. EXAMPLE FOR COUPLED FUNCTIONAL REQUIREMENTS

eters c'_1, c'_2, etc. refer to stud instance two. The center distance between the two studs is D.

In this case, all parameters for each of the two tolerance primitives are now constrained and the primitives are coupled. New constraints for the two zones must then be added to the constraint network to enforce the functional requirements and the coupling condition.

$$0 \leq c_3 \leq 2\arctan\left[\frac{\frac{(s_{N1}+2|a_1|-s_1)}{l_1}}{1+\sqrt{1+\left(\frac{(s_{N1}+2|a_1|+s_1)}{l_1}\right)\left(\frac{(s_{N1}+2|a_1|-s_1)}{l_1}\right)}}\right] \quad (6.2)$$

$$0. \geq 2\sigma + s_1 \sec c'_3 + l_1 \tan c'_3 - s'_{N1} - 2|a'_1| \quad (6.3)$$

$$\sigma = \sqrt{(c'_1 + l_2 \sin c'_3 \cos c'_4 - c'_X)^2 + (c'_2 + l_2 \sin c'_3 \sin c'_4 - c'_Y)^2} \quad (6.4)$$

$$c'_X = \sqrt{D^2 - (c'_Y - c_2)^2} + c_1 \quad (6.5)$$

$$c'_Y = \sqrt{D^2 - (c'_X - c_1)^2} + c_2 \quad (6.6)$$

Further, the design might lead to positions specified for each of the studs according to a three datum system, as shown in detail C of Figure 6.5. The conservative conditional tolerance relationships for this requirement are expressed in Equations 6.7 – 6.15. Parameters c_1, c_2, etc. refer to stud instance one while parameters c'_1, c'_2, etc. refer to stud instance two. The center distance between the two studs is D.

In this case, several parameters are now defined relative to fitted datums. No coupling would be detected here since the functional requirements of detail A and detail B use datum systems and assumptions which are consistent.

In some cases, the values associated with the centerline distance and the positions with respect to the datum system could be inconsistent. In such a case, the invalid portion of the tolerance specification is determined via constraint propagation. The previous coupling could effectively be removed since the Equations 6.2-6.6 and Equations 6.7-6.15 are redundant.

$$0. \geq 2\sigma + s_1 \sec c_3 + l_1 \tan c_3 - s_{N1} - 2|a_1| \quad (6.7)$$

$$\sigma = \sqrt{(c_1 + l_2 \sin c_3 \cos c_4 - c_X)^2 + (c_2 + l_2 \sin c_3 \sin c_4 - c_Y)^2} \quad (6.8)$$

$$c_X = c_{N1} \quad (6.9)$$

$$c_Y = c_{N2} \quad (6.10)$$

$$0. \geq 2\sigma' + s_1 \sec c'_3 + l_1 \tan c'_3 - s'_{N1} - 2|a'_1| \quad (6.11)$$

$$\sigma' = \sqrt{(c'_1 + l_2 \sin c'_3 \cos c'_4 - c'_X)^2 + (c'_2 + l_2 \sin c'_3 \sin c'_4 - c'_Y)^2} \quad (6.12)$$

$$c'_X = c'_{N1} \quad (6.13)$$

$$c'_Y = c'_{N2} \quad (6.14)$$

$$D = \sqrt{(c_X - c'_X)^2 + (c_Y - c'_Y)^2} \quad (6.15)$$

The example illustrates all of the steps that are required to incrementally handle coupled functional requirements. Clearly, it would be very difficult to define general case algorithms for this approach. For well-contained sets of design primitives, however, it should be possible to define a priori all combinations of primitives which are consistent. As well, it may be possible to define consistent combinations of assumptions and datum systems. This second approach would require an extension to the framework for tolerance synthesis since currently there is no provision for storing and reasoning about the assumptions supporting each tolerance primitive.

6.6 Closure

Design synthesis and composition of tolerance requirements have been discussed to motivate the use of composition techniques for tolerance specification. Geometric tolerancing is ambiguous in that a partial solution may represent many different tolerancing strategies. Geometric tolerancing is also nonunique — there are many different tolerance specifications that enforce a particular set of functional requirements. It is often the case, as well, that requirements imposed during tolerance specification are contradictory and conflicts must be resolved before a valid tolerance specification is achieved.

Composition techniques are proposed to provide for the incremental development of tolerance specifications. Cases of independent and coupled functional requirements are defined and approaches are described to support composition for each case. For independent functional requirements, the framework for tolerance synthesis provides for automatic and interactive evaluation of validity for tolerance specifications. Competing requirements can be intersected automatically to maintain validity during the design effort. When requirements are contradictory, the designer is supplied with information about all parameters and equations that contribute to the conflict. For coupled functional requirements, additional checking must be done for each tolerance primitive and further analysis may be required to provide appropriate conditional tolerance relationships. Still, the framework for tolerance synthesis provides the majority of the representation and computing procedures required for these cases.

Chapter 7

Guiding Design and Explanation

Engineering design often proceeds from an incomplete set of requirements and design details. Working from what is already known, additional features are chosen to coincide with features that have been determined. The detailed design representation of our framework for tolerance synthesis maintains knowledge of both design and manufacturing features. Due to the size of the design representation, however, a complete display of the design features is not useful for guiding design efforts. Additional computing is done [82] to provide compact views of the design. These views, or explanations, can also be tailored to communicate design or manufacturing oriented details.

In this chapter, the explanation problem is described and a hierarchical structure is presented for the generation of explanations. A rule-based approach for explanations is then described with examples. Extensions for different types of explanations and optimizations are also discussed.

7.1 The Explanation Problem

The constraint representation employed in the framework for tolerance synthesis maintains a complete description of nominal solids, tolerance parameters, and the conditional tolerance relationships that must be enforced to ensure that functional requirements are satisfied. Chapters 4, 5 and 6 have shown how this representation can be used advantageously to model functional requirements, measure sufficiency, and ensure validity.

A constraint representation was chosen for the framework for tolerance synthesis because geometric tolerancing is ambiguous and nonunique. Each partial solution to a tolerance specification may represent many different tolerancing strategies. As well, there are many different tolerance specifications that enforce a particular set of functional requirements. We propose that improved tolerance specifications can be generated by interpreting partial solutions to tolerance specifications that are represented in the constraint network. This approach should also prove beneficial for other engineering design synthesis tasks.

While many of the tolerance specification tasks can be highly automated, the underlying design representation must be accessible to engineers who are engaged in design activities. Nominally, the design engineer should be able to ask questions of the representation about the design and receive responses that reflect the current state of the detailed tolerance representation. Each question is analogous to a database query and the response is equivalent to an explanation of the tolerance representation. Each response to a query of the constraint network is called an explanation.

Queries of the constraint network must be handled judiciously to ensure that useful information is provided. For small problems, such as the prisms example of Figure 6.3, full explanations generated from the constraint network are easily comprehended. Table 7.1 shows an explanation of how the volume of prism-2 could be calculated. The explanation recounts all of the constraints, as represented in Figure 6.4, that influence the volume parameter. Three different approaches are suggested to arrive at a value for volume. The smallest examples of constraint networks holding tolerance primitives, however, can generate explanations that are very difficult to comprehend. Table 7.2 shows a portion of the textual explanation generated to describe how the orientation parameter c3 can be calculated for an instance of a cylindrical orientation tolerance primitive. A full trace of the explanation and a graphical representation are included in Appendix C.

In evaluating the explanation of Table 7.1, it can be concluded that four parameters are currently unbound and that three different combinations of parameter assignments can be used to arrive at a value for volume. If a designer is concerned only with the parameters that should be set, a more compact listing of parameter combinations would be more useful. All of the details concerning calculation operations and assignments can be removed. This is an example of aggregation. Intermediate details are removed or transformed to provide a more compact view of the design.

In evaluating the explanation of Tables C.1 – C.6, a variety of conclusions can be drawn. If the designer is concerned only with the relationships influenced by tolerance assertion parameters, the explanation can be pruned of equations related to the nominal shape of the part. This is an example of segregation. Parts of the explanation are removed to focus on only one aspect of the design such as manufacturing requirements.

The prism and tolerance primitive examples demonstrate that the constraint network can be used to explain *how* a parameter might be calculated when no value is currently assigned. The detailed representation used in the framework for tolerance synthesis also allows for explanations of *why* particular values have been assigned. This type of explanation is particularly important when several tolerance primitives are included in a design. The *why* explanations describe the order of calculations and effectively describe unique conditions for each tolerance primitive parameter.

Table 7.1: Trace of Prism Example Constraint Explanation

```
PRISM-2 VOLUME
  #<SLOT: VOLUME > DETERMINED-BY
    SETTING  #<SLOT: VOLUME >
    #<SLOT: VOLUME > COULD-BE-CALCULATED-WITH
     <#:*6780:MULTIPLIER>  USING
      *
        M1
          SETTING  #<SLOT: AREA >
          #<SLOT: AREA > COULD-BE-CALCULATED-WITH
           <#:*6779:MULTIPLIER>  USING
            *
              M1
                SETTING  #<SLOT: LENGTH >
              M2
                SETTING  #<SLOT: HEIGHT >
          M2  VALUE  #<SLOT: CELL-326 >  [1.0,1.0]
that's all
```

7.2 Structure for Explanation Task

Explanations for constraint network queries are generated by traversing the graph of objects which are associated with the constraint network. The framework for tolerance synthesis uses IDEEA constraint networks [23] which include three different types of objects. Cell objects are associated with each parameter value. Constraint objects are associated with each equation. Repository objects are associated with each collection of cells that share the same value. For example, when two parameter values are equated with a constraint, the cells associated with each parameter are assigned to one common repository. In addition the truth maintenance system employed by IDEEA builds a network of the justifications and assumptions that support each cell and constraint.

The verbose explanations shown in Tables 7.1 and 7.2 are the result of simple constraint network traversals. The graph traversal stops when a terminal leaf is found or when a cycle is detected. Cycles are detected by uniquely marking nodes in the graph as the traversal continues. Initially a list structure is built up for the explanation and some rudimentary formatting is then done to produce the explanation.

Each constraint network query begins as either a *how* or *why* question. When a query is modified to achieve segregation or aggregation, the explanation is further pruned. Figure 7.1 shows the hierarchical structure defined for the explanation task.

Table 7.2: Partial Trace of Tolerance Primitive Constraint Explanation

```
#<SLOT: C3 >  DETERMINED-BY
SETTING  #<SLOT: C3 >
#<SLOT: C3 >  COULD-BE-CALCULATED-WITH
<#:*3725:MULTIPLIER>  USING  *
M1  ALWAYS  #<SLOT: CELL-371 >  2.0
M2
#<SLOT: CELL-372 >  COULD-BE-CALCULATED-WITH
<#:ATAN3724:TAN>  USING  ATAN
TAN+1
#<SLOT: CELL-369 >  COULD-BE-CALCULATED-WITH
<#:/3723:MULTIPLIER>  USING  /
PRODUCT
#<SLOT: CELL-365 >  COULD-BE-CALCULATED-WITH
<#:/3708:MULTIPLIER>  USING  /
PRODUCT
#<SLOT: CELL-323 >  COULD-BE-CALCULATED-WITH
<#:|-3707|:ADDER>  USING  -
SUM
#<SLOT: CELL-320 >  COULD-BE-CALCULATED-WITH
<#:|+3706|:ADDER>  USING  +
X  #<SLOT: CELL-318 >  DETERMINED-BY
SETTING  #<SLOT: SN1 >
#<SLOT: SN1 >  COULD-BE-CALCULATED-WITH
<#:TOLPRIM-SO1.M7685:TOLPRIM-SO1.P>  USING  SN1
A1  #<SLOT: CELL-307 >  DETERMINED-BY
SETTING  #<SLOT: A1 >
#<SLOT: A1 >  COULD-BE-CALCULATED-WITH
<#:ABS3704:ABSOLUTE>  USING  ABS
PNUM  #<SLOT: CELL-313 >  DETERMINED-BY
#<SLOT: CELL-316 >  COULD-BE-CALCULATED-WITH
<#:*3705:MULTIPLIER>  USING  /
PRODUCT  #<SLOT: CELL-314 >  DETERMINED-BY
#<SLOT: CELL-319 >  WHICH-CYCLES
M1  ALWAYS  #<SLOT: CELL-315 >  2.0
#<SLOT: A1 >  COULD-BE-CALCULATED-WITH
<#:ABS3709:ABSOLUTE>  USING
ABS
```

7.2. STRUCTURE FOR EXPLANATION TASK

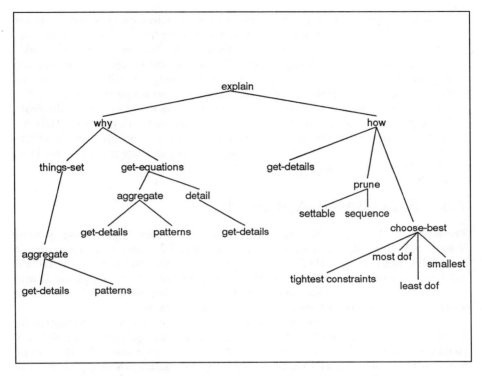

Figure 7.1: Hierarchical Structure for Explanation Task

The top two levels of the hierarchy, corresponding to control knowledge about the explanation task, are treated as domain independent levels. The lower levels of the hierarchy correspond to domain knowledge and may be tailored for each particular application.

A *why* query involves first gathering the set of cells that contribute to the value being queried. This is easily done by using the truth maintenance functions of IDEEA. The equations behind each cell are then explained by traversing the constraint network. The resulting explanation can then be pruned according to user preferences for detail and particular patterns. An example of this, using rules, is shown later for the *how* query. A *how* query involves first generating the detailed explanation through graph traversal and then pruning and selecting according to user preferences. In many cases, these two types of queries are intermingled. If some parameters in the constraint network are unbound, a *why* query will include the results of smaller *how* queries. As well, a *how* query can ultimately include results from smaller *why* queries for values that are bound.

This hierarchy was chosen for the framework for tolerance synthesis to ensure that control knowledge is separated from domain knowledge. The control knowledge remains the same for a variety of design tasks while the domain knowledge is varied for different parts, users, or design phases. The control knowledge describes the way that explanations are arrived at. The domain knowledge describes conditions for useful explanations and methods for aggregating and segregating relations.

In some cases, domain knowledge can be used to effectively limit both the depth and breadth of the search that results during constraint network traversal. More importantly, the separation of control and domain knowledge allows common algorithms to be reused among many different explanation strategies while providing flexibility to tailor the results for particular applications.

7.3 Rule-Based Approach

The hierarchical approach illustrated in Figure 7.1 has been implemented by using IDEEA and its facility for rule-based reasoning. Detailed explanations are generated using LISP functions and then pruned by evaluating rules and facts in the IDEEA database. For **why** explanations, **things-set** and **get-equations** are implemented with lisp functions. For **how** explanations, **get-details** is implemented with lisp functions while **prune** and **choose-best** are implemented largely with rules.

The lisp function for **things-set** accesses the truth maintenance information for the value that is being queried. For **why** explanations, the **get-equations** task is implemented using the functions shown in Table 7.3. The constraint network traversal begins with a value cell and continues with the constraints which justify it. The traversal progresses through constraints to the other values associated with each constraint and continues until assumptions such as user-set values are reached. For **how** explanations, the **get-equations** task is implemented using the top level function shown in Table 7.4 and several supporting functions which build up the explanation list structure. A similar traversal is done through value cells, repositories, and constraints.

IDEEA rules are used to implement **prune** and **choose best**. The verbose explanation generated via **get-equations** is first instantiated in the IDEEA database and then reduced via rule-based reasoning. Facts defined for each application are used to guide the rule-based pruning. Some of the facts used for the prism example are shown in Table 7.5. Forms such as (**settable area**) guarantee that explanations will only include parameters that are useful for setting. Forms such as (**calculating length choose least-dofs**) describe a user preference for the way that an explanation should be generated.

For example, Table 7.6 shows an aggregated explanation generated for the same query explained in Table 7.1. The verbose explanation is first reduced to two parallel explanations. Each explanation is evaluated to ensure that it includes parameters that may be set. One explanation has to do with setting the **volume** parameter while

7.3. RULE-BASED APPROACH

Table 7.3: Lisp Functions for Why Explanation

```
(defun chase—vja—parents (something &aux newthing)
  "function chase-vja-parents expects a value, assumption or
   justification object and returns a list with the object and
   its parents in which will be objects with justifications,
   nil returned if wrong object  i.e., no parents"

  (when something
    (cond
      ((listp something)
        (for (x :in something)
             :filter (chase—vja—parents x)))
      (t
        (setq newthing (vja.parents something))
        (when newthing (cons (explain—label something)
                             (chase—vja—parents newthing)))))))

(defun vja.parents (something)
  "function vja.parents expects an object of value, assumption
or justification and returns a list of parents - as represented in
each object ,  nil if unacceptable object"

  (cond ((v—p something)
         (v—justifications something))
        ((tms.justification? something)
         (tms.justification—antecedents something))
        ((tms.assumption? something)
         (tms.assumption—environments something))
        (t nil)))
```

Table 7.4: Lisp Functions for How Explanation

(defun to—calculate (something &key (ultimately nil) (mark nil) last—rep)

  "function to-calculate expects an object of cell, and returns
a list of equations that could be used to determine the object
- each equation also includes objects which in turn may be
determined by other objects - the list stops on values or cells
without owners frame-slots and cycles. Cycles are described one
level deeper than where the traversal stopped. nil is returned
if no equations or wrong object.
:ultimately set to true results in dependencies
 being included for values
:mark and :last-rep are used for recursive calls."

  (unless mark (setq mark (gensym)))

  (cond

    ((dirty—markp something mark) nil)
    ((cell—p something)
     (mark—dirty something mark)
     (get—eq—cell something :ultimately ultimately :mark mark))

    ((rep—p something)
     (mark—dirty something mark)
     (get—eq—rep something :ultimately ultimately :mark mark))

    ((con—p something)
     (mark—dirty something mark)
     (get—eq—con something :ultimately ultimately
           :mark mark :last—rep last—rep))

    (t nil)))

7.3. RULE-BASED APPROACH

Table 7.5: IDEEA Database Facts for Prism Explanation

```
(defdata cascade—data
  :facts
  ;; static facts for the prisms example

  (settable area)
  (settable volume)
  (settable height)
  (settable width)
  (settable length)

  (calculating area choose tightest—values)
  (calculating volume choose tightest—values)
  (calculating height choose loosest—values)
  (calculating width choose loosest—values)
  (calculating length choose loosest—values)

  (calculating area choose most—dofs)
  (calculating volume choose most—dofs)
  (calculating height choose least—dofs)
  (calculating width choose least—dofs)
  (calculating length choose least—dofs)

  ;; additional facts omitted ...

  )
```

Table 7.6: Aggregated Explanation for Prism Explanation

```
to calculate, #<SLOT: VOLUME >
set   HEIGHT   of frame   PRISM-2
set   LENGTH   of frame   PRISM-2

to calculate, #<SLOT: VOLUME >
set   VOLUME   of frame   PRISM-2

end of explanation
```

the other has to do with calculating a value for **volume**. The explanation of the calculation requires that degrees of freedom be evaluated and then used to choose a particular explanation. In this case, the most degrees of freedom is used as a criteria for the explanation. This criteria produces the first explanation in Table 7.6. Note that the possibility of setting the **area** parameter is omitted since it involves fewer degrees of freedom. The rules that were used to generate this explanation are shown in Table 7.7.

This approach to explanation is quite flexible but requires significant computing resource. The generation of each verbose calculation can, in the worst-case, generate a list that displaces as much memory as the initial constraint network. The determination of terminal nodes in the explanation tree and the evaluation of attributes such as degrees of freedom can result in an exhaustive search of the explanation list structure.

To a large degree, the constraint network representation includes enough information to reduce the complexity of generating explanations. The marking functions used to generate verbose explanations could be employed to add state information to the constraint network. In a more robust rule-based implementation, it would be possible to evaluate the constraint network and the explanation state directly via rule evaluation. The use of backward chaining rules would dramatically reduce the complexity of determining terminal nodes and attributes such as degrees of freedom.

7.4 Extensions for Optimization Formulations

The verbose explanations generated from the constraint representation also provide all of the information that is required to generate optimization formulations for the calculation of unbound values. By performing a series of substitutions on the explanation, it is possible to generate a system of equations that will lead to a value for the parameter that is explained. The approach required for this transformation is summarized in Algorithm 7.1. A canonical variable list for the explanation is first

7.4. EXTENSIONS FOR OPTIMIZATION FORMULATIONS

Table 7.7: IDEEA Rules for Prism Explanation

```
(defrule determined-by
  (IF (?x DETERMINED-BY ?y))
  (THEN (dolist (explanation ?y) (lassert explanation))))

(defrule name-of-value
  (IF (setting ?x) TEST (slot-p ?x))
  (THEN (lassert '( ?x ,(slot-name ?x)))))

(defrule set-value
  (IF (SETTING ?x) (?x ?y) (SETTABLE ?y))
  (THEN (print-explain ?x (list 'SETTING ?x))))

(defrule could-be-calculated
  (IF (?x COULD-BE-CALCULATED-WITH ?y))
  (THEN (lassert-values-determining ?x ?y)))

(defrule good-determining-values
  (IF (determining ?x ?y))
  (THEN (lassert '(DOFS ,(count-setting-leaves ?y) ?x ?y ))
        (lassert '(VALS ,(count-value-leaves ?y) ?x ?y ))
        (lassert '(PINS ,(count-leaves ?y) ?x ?y))))

(defrule get-least-dof-0
  (IF (determining ?x ?explain) (DOFS ?count ?x ?explain)
      (not (LEAST DOFS ?x ?leastcount)))
  (THEN (lassert '(LEAST DOFS ?x ?count))))

(defrule get-least-dof-1
  (IF (DETERMINING ?x ?explain) (DOFS ?count ?x ?explain)
      (LEAST DOFS ?x ?leastcount) -> ?rem
      TEST (< ?count ?leastcount))
  (THEN (retract ?rem) (lassert '(LEAST DOFS ?x ?count))))

(defrule get-most-dof-0
  (IF (determining ?x ?explain) (DOFS ?count ?x ?explain)
      (not (MOST DOFS ?x ?mostcount)))
  (THEN (lassert '(MOST DOFS ?x ?count))))
```

generated along with a set of formulas from the constraints and cells of the constraint network. The canonical variable names and then the formulas are substituted to build up a system of equations. The resulting system of equations may include interval values which must be interpreted further to yield inequality conditions for the system of equations.

Algorithm 7.1 (Optimization Formulation From Constrain Network)
  Step 1. Let *Explanation* be the verbose explanation.
  Step 2. Generate *UniqueVars* as an association list of canonical names for each value cell in the explanation.
  Step 3. Generate *Formulas* by reducing *Explanation* to a set of formulas of the form $X1 = X33 * X54$.
  Step 4. Reduce *Formulas* by eliminating forms where a constraint-owned cell is equated to a frame-slot. In this case, replace references to the constraint-owned cell with the associated frame-slot.
  Step 5. Reduce *Formulas* by eliminating forms where a constraint-owned cell is equated to a bound value. In this case, replace references to the constraint-owned cell with the bound value.
  Step 6. Let *Combinations* be the power set of *Formulas*.
  Step 7. Reduce each member of *Combinations* by performing back substitution.
  Step 8. Return each member of *Combinations*, along with *UniqueVars*, as an optimization formulation.

For example, Table 7.8 shows the intermediate results when Algorithm 7.1 is applied to the verbose explanation of Table 7.1. Note that this result is quite different from a simple restatement of the original constraint network. Each bound value in the constraint network reduces the number of equations found in the optimization formulation.

7.5 Closure

The detailed design representation of the framework for tolerance synthesis maintains knowledge of both design and manufacturing features. This representation and its implementation in a constraint network specifically address the problems of ambiguity and uniqueness that are inherent in geometric tolerancing. The size of the design representation, however, is a burden when designers wish to ask questions about the current state of a design.

An approach for the explanation of constraint networks is proposed as a means of providing compact and comprehensible views of the design representation. A hierarchical structure, where control and domain knowledge are separate, is employed

7.5. CLOSURE

Table 7.8: Intermediate Results for Optimization Formulation

Step 2.

((#<SLOT: VOLUME > . X1) (#<SLOT: AREA > . X2)
 (#<SLOT: LENGTH > . X3) (#<SLOT: HEIGHT > . X4)
 (#<SLOT: CELL−414 > . X5))

Step 3.

((X1 = X2 * X5) (X5 = [1.0,1.0]) (X2 = (* X3 X4)))

Step 4. (NA)

Step 5.

((X1 = X2 * [1.0,1.0]) (X2 = (* X3 X4)))

Step 6.

(((X2 = (* X3 X4)) (X1 = X2 * [1.0,1.0])))

this is trivial case for a case where there are 2 definitions such as ((X1 = X6) (X1 = (X2 * [1.0,1.0])) (X2 = (* X3 X4))))

the power set is

(((X2 = (* X3 X4)) (X1 = (X2 * [1.0,1.0])))
 ((X2 = (* X3 X4)) (X1 = X6)))

Step 7.

X1 = ((* X3 X4) * [1.0,1.0])

Step 8.

X1 = ((* X3 X4) * [1.0,1.0])
((#<SLOT: VOLUME > . X1) (#<SLOT: AREA > . X2)
 (#<SLOT: LENGTH > . X3) (#<SLOT: HEIGHT > . X4)
 (#<SLOT: CELL−414 > . X5))

so that explanations may be tailored to particular domains. Further, the representation employed for the framework for tolerance synthesis is shown to be effective for the generation of optimization formulations which can be used for final synthesis of tolerance specifications.

Chapter 8
CASCADE-T Implementation

Previous computer-based approaches to tolerance specification have focused on representation and analysis, i.e. how tolerance information is stored and whether specified tolerances satisfy the design requirements. To provide for tolerance synthesis, we have developed a framework for tolerance synthesis, as described in Chapter 3, and proposed that the interaction and integration among computing tasks and representation contribute most significantly to tolerance synthesis. Using the techniques described in Chapters 4 – 7, the framework for tolerance synthesis can be used to extend current CAD systems.

To demonstrate the improved facilities for tolerance specification which result from the integration and interaction within the framework, a prototype computing environment has been developed. The computing environment is called CASCADE-T : Concurrent, computer-Automated methods for the Synthesis of Competing Design Elements and geometric Tolerances. Synthesis is emphasized because CASCADE-T allows tolerance specifications to be generated and checked automatically during the design process.

The CASCADE-T implementation demonstrates how the methods developed in this book can be used efficiently for tolerance specification. A very detailed and consistent representation is maintained for relationships between functional requirements and geometry. Incremental reasoning about tolerance requirements for geometric shapes or parts in mechanical assemblies is supported. CASCADE-T is integrated with the BRL solid modeler [45, 44] while remaining independent of any particular modeling scheme. In this chapter, a design scenario is described to show where CASCADE-T functions fit into the design process. The current implementation is briefly described and several examples are shown.

8.1 Design Scenario

Using a CAD system extended by the framework for tolerance synthesis, a designer should be able to first describe a part or assembly of parts using primitive solids provided by the CAD system. Working with the nominal description of the part, the designer should then be able to specify relationships between features by using a library of tolerance primitives. The elements in the library of tolerance primitives would correspond to different kinds of shapes and functional requirements. For each tolerance primitive, appropriate features and datums would be specified by the designer so that the features of each solid are connected with the parameters of tolerance primitives. As tolerance primitives are instanced and associated with solids, governing equations and parameters are added to the constraint network and the effects are propagated.

During a design session, a CAD system extended by CASCADE-T would support tolerance specification by monitoring the design state and responding to user queries. As primitives are instanced, coupling conditions would automatically be recognized and handled in the constraint network. Validity and sufficiency would be monitored as the design progresses. The designer would be able to query for insufficiently constrained features and suggestions for accomplishing appropriate constraint - the system would respond with useful explanations suited to the particular application. When a particular functional requirement implies a perfect form or position, the system should signal a contradiction and suggest different combinations of features that might be modified to correct the problem.

8.2 Description of Implementation

The CASCADE-T implementation provides many of the features described in the preceding scenario. A library of tolerance primitives has been implemented by using the Intelligent Design Environment for Engineering Automation (IDEEA) and its constraint definition facility. Following the Application Interface Specification (AIS) [50, 51], an interface between a solid modeler and the IDEEA environment has been implemented. Once a part is defined in the solid modeler, the IDEEA environment queries the solid modeler for a description of the part. Constraints and parameters associated with tolerance primitives may be connected with the parameters which describe each primitive solid. Constraint propagation allows for the validity of tolerance specifications to be checked each time additions or modifications are made to the design. Contradictions are signaled to the user and explanations of the constraint network may be generated to suggest design changes that will be consistent with the functional requirements described by the tolerance primitives.

8.2. DESCRIPTION OF IMPLEMENTATION

CASCADE-T is implemented with IDEEA which provides constraint networks, frames, truth-maintenance, and rule-based reasoning within a COMMONLISP programming environment. The constraint networks of IDEEA are similar to those described by Steele [64]. A much richer set of values, however, are supported for IDEEA computations. Scalar values, symbols, intervals, and sets may all be computed within IDEEA constraint networks. CASCADE-T, in particular, largely uses interval calculations. The frame system of IDEEA is similar to one described by Charniak [5] and includes additional features for multiple inheritance and pattern matching. In IDEEA, values are implemented as objects and common calculation operators can be repeatedly defined to accommodate calculations for different combinations of objects. The state of each value and each constraint is also accounted for through the use of a truth-maintenance system which keeps a detailed account of dependencies which exist among constraints and values. The rule-based system in IDEEA is similar to the CLIPS program [46] and allows all frame instances to be accessed during rule executions.

As described in Chapter 4, the tolerance primitives are defined using the IDEEA constraint language. The definitions for the assembly and material bulk tolerance primitives are also included in Appendix A. The tolerance primitives use a number of constraint definitions and macros that are extensions to the IDEEA implementation. Solid models are built up using the BRL solid modeler and then associated with tolerance primitives by using CLOS-AIS [83], an AIS interface between BRL and the LISP environment. As validity is evaluated via constraint propagation, newly calculated values may change the nominal geometry described within the solid modeler. Currently, updates from the solid modeler to CASCADE-T are only performed when a model file is opened.

Typical steps for opening a model file are shown in Table 8.1 and the resulting display of the model is shown in Figure 8.1. Once a model has been opened via the AIS interface, all primitive solids and associated parameters may be accessed via the IDEEA frame system. Table 8.2 shows a typical constraint language form that is used to associate a tolerance primitive with a primitive solid in the solid modeler.

Figure 8.2 shows a typical display from CASCADE-T. Frame editor windows, such as the **STUD-1 Editor Window** are available to edit the parameters of tolerance primitives and solids. When contradictory values or ranges of values are introduced by the designer, a contradiction is signaled as shown in Figure 8.2 and the user is able to choose parameters which should be overruled or constraints that should be ignored. In the background of Figure 8.2, a graph of the constraint network is also displayed. Similar graphs may be generated for more detailed views of the network and for dependencies.

Table 8.1: Open a Model File in CASCADE-T

```
> (setf example-1 (clos:make-instance
                   'ais:brl-modeler :current-directory "~/rockwell/brlwork"
                   :display-site "silicon:0.0" :computing-site "carbon"))
#<Brl-Modeler #X184E3C6>

> (ais:get-model example-1 (merge-pathnames
                             (make-pathname :name "thesis-ex1.g")
                             "~/rockwell/brlwork/"))
#P"/home2/wilhelm/rockwell/brlwork/thesis-ex1.g"

> (ais:start-modeler example-1)
"BRL-CAD Release 3.5 Graphics Editor (MGED) Compilation 74
    Fri Apr 12 18:29:39 PDT 1991
    wilhelm@carbon:/home/wilhelm/rockwell/brlcad/bw-mged

ATTACHING nu (Null Display)
Untitled MGED Database (units=mm)
mged> "
0
"modeler started"

> (ais:brl-attach-X example-1)
"X Display [silicon:0.0]? Font [6x10]?
 ATTACHING X (X Window System (X11))"

> (brl-prims-to-ideea example-1)
instantiating "/root/ex-base-1/base-1"
instantiating "/root/ex-base-1/stud-1"
instantiating "/root/ex-top-1/top-1"
instantiating "/root/ex-top-1/hole-1"
no ideea frame instantiated for #<3d-Vector #X17E3D76>
no ideea frame instantiated for #<3d-Point #X17E3BFE>
no ideea frame instantiated for #<3d-Point #X17D1E66>
no ideea frame instantiated for #<3d-Vector #X1A06FAE>
no ideea frame instantiated for #<3d-Point #X1A0701E>
no ideea frame instantiated for #<3d-Point #X1A090DE>

> (ais:brl-evaluate-all example-1)
```

8.2. DESCRIPTION OF IMPLEMENTATION

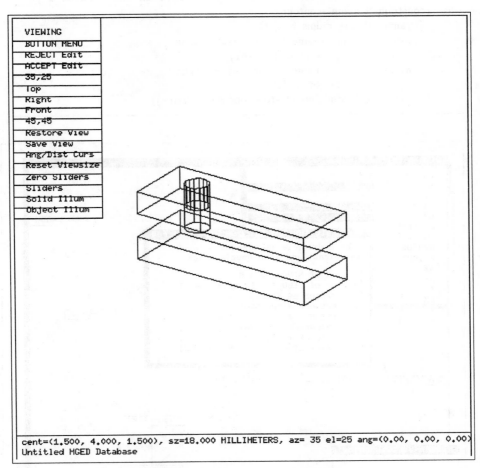

Figure 8.1: BRL Display for Model File in CASCADE-T

Table 8.2: Example Constraints for Model File in CASCADE-T

```
(with-binding-contour
   ((frame (ideea-solid-frame-for-pathname "/root/ex-base-1/stud-1")))
   (constrain-binding-contour
    ((frame  l1 sn1 radius height))
     (with-constraint-name "tp-to-ais-radius"
        (constrain  (2 * radius) = sn1))
     (with-constraint-name "tp-to-ais-height"
        (constrain  height = l1)))
    (assembly-cyl-stud-orient-tol-constraints frame))
```

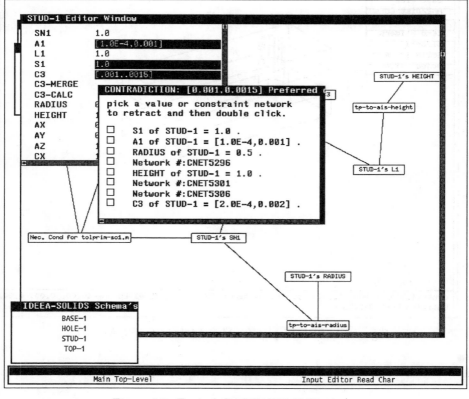

Figure 8.2: Typical CASCADE-T Display

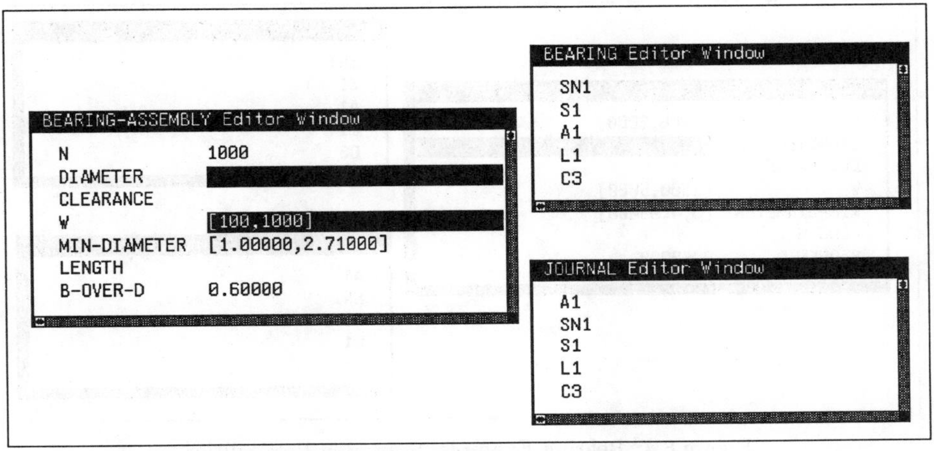

8.3. EXAMPLES   91

```
┌─────────────────────────────────────────────────────────────────────┐
│                                          ┌─BEARING Editor Window─┐  │
│                                          │  SN1                  │  │
│                                          │  S1                   │  │
│   ┌─BEARING-ASSEMBLY Editor Window─┐     │  A1                   │  │
│   │  N            1000             │     │  L1                   │  │
│   │  DIAMETER                      │     │  C3                   │  │
│   │  CLEARANCE                     │     └───────────────────────┘  │
│   │  W            [100,1000]       │                                │
│   │  MIN-DIAMETER [1.00000,2.71000]│     ┌─JOURNAL Editor Window─┐  │
│   │  LENGTH                        │     │  A1                   │  │
│   │  B-OVER-D     0.60000          │     │  SN1                  │  │
│   └────────────────────────────────┘     │  S1                   │  │
│                                          │  L1                   │  │
│                                          │  C3                   │  │
│                                          └───────────────────────┘  │
└─────────────────────────────────────────────────────────────────────┘
```

Figure 8.3: Bearing Example, Initial Problem

8.3 Examples

To demonstrate the features of **CASCADE-T**, a bearing design problem where ranges of design parameters are considered is first presented. This type of approach is usually associated with earlier stages of a design. The example shows how the details embedded in **CASCADE-T** tolerance primitives can be used to guide a design.

Figure 8.3 shows the variables involved in this problem. For the bearing-assembly, N and W refer to rotation speed and loading. MIN-DIAMETER refers to the minimum diameter allowed by the Walsh design method. DIAMETER is the journal diameter used for the design and LENGTH is the journal length. B-OVER-D refers to the length to diameter ratio.

For the bearing and journal, cylindrical fit parameters include the nominal diameter for the cylinder, SN1, the allowable variation of a perfect form cylinder, S1, the allowable variation of a perfect form cylinder, A1, the measured length of the cylinder, L1, and the attitude or orientation angle of the cylinder about the x or y axis, C3.

When N and W are specified, a minimum diameter is calculated and shown in the bearing-assembly window.

CHAPTER 8. CASCADE-T IMPLEMENTATION

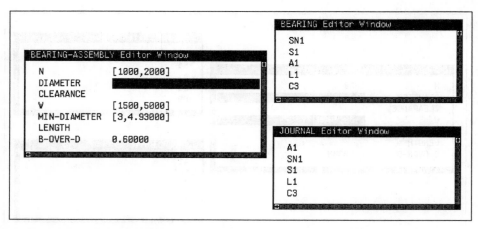

Figure 8.4: Bearing Example, Searching Possibilities

Figure 8.4 demonstrates one of the additional features of CASCADE-T - by specifying a range of diameters and rotation speeds, a set of plausible loads is identified. Note that the order and type of query influence the values that are presented - there are other load/rpm combinations that result in diameters of 3 to 5 inches. This query determines loads for the diameters and rpms specified.

The ranges that are generated for the bearing parameters are bounded by the worst-case condition, the highest load for example, and the minimum condition. The lower bound suggests the smallest value that could be used if other ranges were reduced. The upper bound suggests the worst-case for the ranges under consideration.

Figure 8.5 shows results for a diameter chosen very close to the minimum required for bearing performance. As shown in the bearing editor window, a very small negative value is calculated for the lower bound of the orientation angle C3. This solution is infeasible as it violates the assembly requirements.

The negative orientation angle is removed by increasing the bearing diameter as shown in Figure 8.6. The cylindrical fit parameters shown in the bearing and journal windows describe the variation allowed for each cylinder.

8.3. EXAMPLES

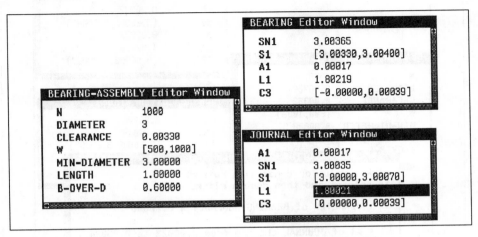

Figure 8.5: Bearing Example, First Solution Infeasible

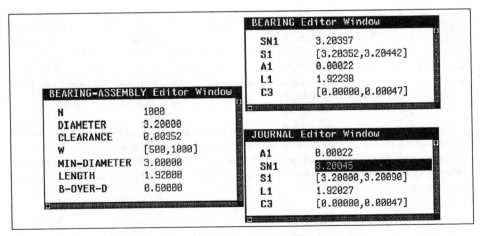

Figure 8.6: Bearing Example, Feasible Solution

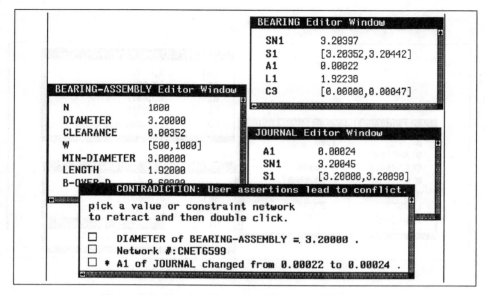

Figure 8.7: Bearing Example, Changing Tolerances

In addressing manufacturing concerns, a larger tolerance on the cylindrical features might be sought as shown in Figure 8.7. A value of 0.00024 is entered for the A1 value of the journal cylinder. This corresponds to a 0.001 unilateral tolerance on the diameter rather than a .0009 tolerance which was previously calculated. This entry results in a contradiction between the values for DIAMETER and A1 and one of the equations that relates these values.

In this case, increasing the diameter substantially will increase the size of the tolerance as shown in Figure 8.8. Equally important, now that such an increase is being considered, other aspects of the design such as load and rotation speed may be relaxed or changed. A query for other loads feasible with the 5.5 inch diameter might suggest ways to make the design more robust.

In stepping through this bearing design example, several different stages of the calculations have been presented to show how tolerance primitives are checked interactively for consistency as the design progresses. As well, the interactive checks guide the engineer to choose design parameters that satisfy the functional requirements for the design.

8.3. EXAMPLES

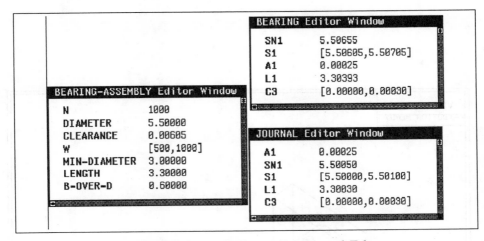

Figure 8.8: Bearing Example, Larger Bearing and Tolerances

The example also shows how CASCADE-T allows details about tolerances, geometric form, manufacturing requirements and other design parameters to be considered very early in the design. When process requirements favor coarser tolerances, CASCADE-T can be used to determine if the new tolerance requirement is feasible and, if not, how to modify the design to satisfy the process requirements.

In a second example, the assembly of a three piece assembly, as shown in Figure 8.9, is considered. A shaft is fitted into two prismatic pieces. In the solid model representation, the left-most prism of Figure 8.9 is labeled base and the right-most prism is labeled plate, as shown in Figure 8.10.

There are two orientation requirements for this assembly. One end of the shaft, labeled shaft-two, must fit in the hole of the base. The other end of the shaft, labeled shaft-one, must fit in the hole of the plate. Table 8.3 shows the tolerance primitives that are specified for these assembly requirements. Additional constraints have been added so that competing requirements for the two orientation angles are merged automatically when an intersection exists.

The solid model parameters of the assembly are connected to the parameters of the tolerance primitives by using constraints and the CLOS-AIS interface. Figure 8.11 shows the growth on shaft-one that results when the radius is changed from 0.5 to 3.0. The tolerance primitives may change the nominal dimensions in the same way although the variations are usually much smaller.

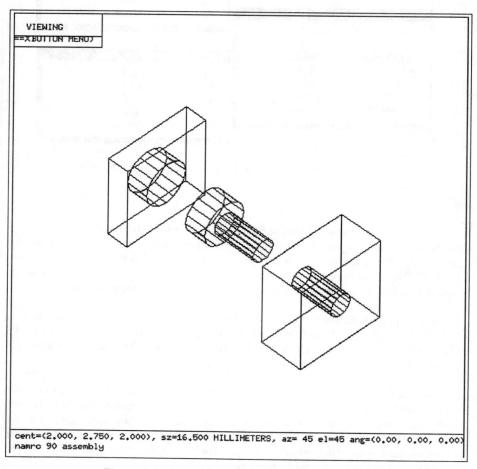

Figure 8.9: Assembly of Base, Shaft, and Plate

8.3. EXAMPLES

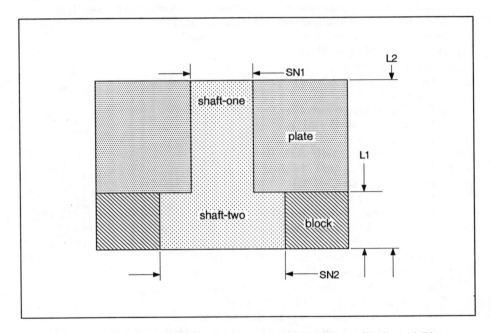

Figure 8.10: Labels and Dimensions for Assembly of Base, Shaft, and Plate

Table 8.3: Tolerance Primitives for Assembly Requirements

```
(with−binding−contour
     ((stud1 (ideea−solid−frame−for−pathname "/shaft/shaft-one"))
      (stud2 (ideea−solid−frame−for−pathname "/shaft/shaft-two")))
    (assembly−cyl−stud−orient−tol−constraints stud1)
    (assembly−cyl−stud−orient−tol−constraints stud2)
    (constrain−binding−contour
     ((stud1 (c3 . stud1−c3))
      (stud1 (c3−merge . stud1−c3−merge))
      (stud2 (c3 . stud2−c3))
      (stud2 (c3−merge . stud2−c3−merge))
      (stud1 (sn1 . stud1−sn1))
      (stud1 (l1 . stud1−l1))
      (stud1 (radius . stud1−radius))
      (stud1 (height . stud1−height))
      (stud2 (sn1 . stud2−sn1))
      (stud2 (l1 . stud2−l1))
      (stud2 (radius . stud2−radius))
      (stud2 (height . stud2−height)))

     (with−constraint−name "tp-to-ais-radius"
        (constrain  (2 * stud1−radius) = stud1−sn1))
     (with−constraint−name "tp-to-ais-height"
        (constrain  stud1−height = stud1−l1))
     (with−constraint−name "tp-to-ais-radius"
        (constrain  (2 * stud2−radius) = stud2−sn1))
     (with−constraint−name "tp-to-ais-height"
        (constrain  stud2−height = stud2−l1))

     (constrain stud1−c3−merge = stud1−c3 merge stud2−c3)
     (constrain stud2−c3−merge = stud1−c3−merge)))
```

8.3. EXAMPLES

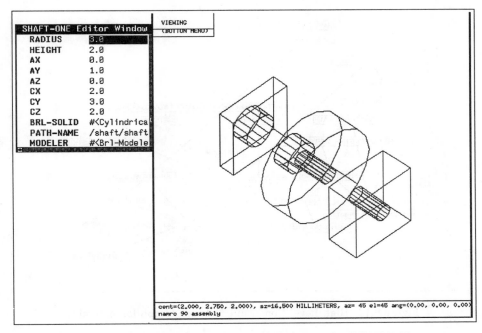

Figure 8.11: Variation of the Solid Model From CASCADE-T

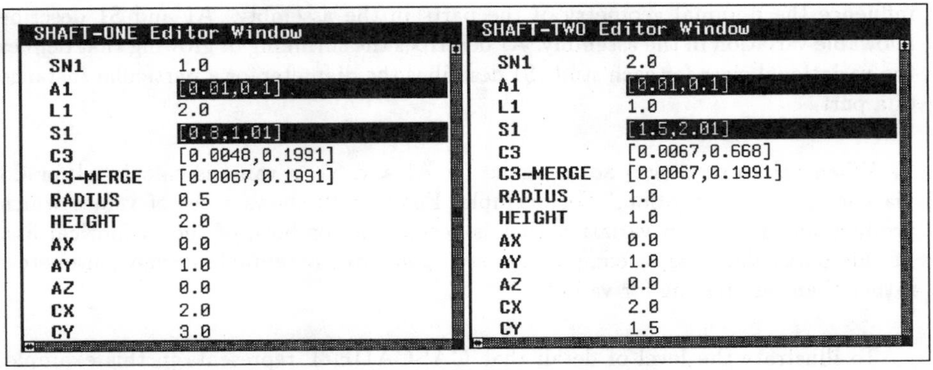

Figure 8.12: Composition of Assembly Requirements in CASCADE-T

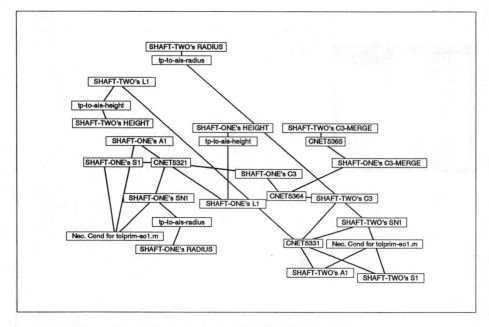

Figure 8.13: High-Level Constraint Network Graph for Assembly

In this assembly, SN1 and L1 are tolerance assembly parameters that directly influence the nominal geometry of the parts in the assembly. A1 and S1 describe allowable variation in the assembly. A1 describes the shrinking or growing that defines the variational class for each stud. S1 describes the diameter for a particular instance of a part.

When values or ranges are assigned for A1 and S1, a value is calculated for C3 via constraint propagation. For example, Figure 8.12 shows a set of values which produce an orientation variation that is acceptable for both of the cylindrical fits. In this particular case, a composition step generates acceptable design parameters rather than just testing for validity.

To illustrate the level of detail that CASCADE-T represents in this example, two views of the constraint network are shown in Figures 8.13 and 8.14. A high-level graph of the constraint network which may be useful for guiding designers is shown in Figure 8.13. A more detailed graph of the constraint network is shown in Figure 8.14.

8.3. EXAMPLES

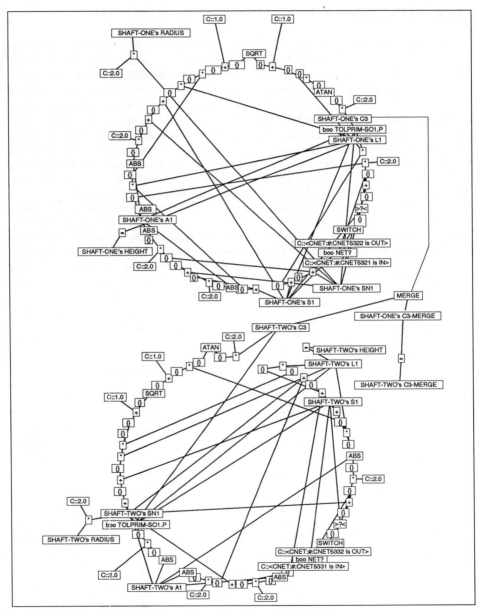

Figure 8.14: Detailed Constraint Network Graph for Assembly

8.4 Closure

The CASCADE-T system represents a significant step in the efficient use of a complex and consistent design representation for tolerance specification. With further development, additional enhancements will exploit this detailed representation. Functional requirements and nominal solid geometry can both be linked to the detailed tolerance representation. In the preceding scenario, it was suggested that a designer participate actively to specify tolerance relationships by choosing many details : features for tolerancing, appropriate datums, and connecting relations. For particular combinations of geometric features and tolerance primitives, it is expected that parts of this task can be assisted or handled by mechanized knowledge sources operating on functional requirements and nominal solid geometry. In addition to merging constraints, the constraint network can be used to enforce validity requirements. As well, the detailed representation can be used for incrementally measuring sufficiency. Finally, feasible ranges for tolerance variables can be further constrained with optimization techniques to yield the best tolerance specifications. Though focused on synthesis, this system also supports analysis; as a part and its features are composed, a tolerance specification is generated that may then be compared with prespecified requirements.

Chapter 9

Conclusions

This research is principally concerned with theory and computer-based tools for the generation of tolerance specifications. Improved methods for tolerance specification will help engineers to design parts with complete and consistent specifications. Successful application of this research to engineering practice promises to reduce costs and errors during design and manufacturing while providing additional opportunities for cooperation among design and manufacturing engineering tasks.

In this book, an approach to geometric tolerances has been developed where function and geometric form are considered simultaneously. A framework of seven interconnected tasks was defined to support both analysis and synthesis of tolerance specifications (Chapter 3). Tolerance primitives, based on a sound theory of tolerancing, were used to represent tolerance relationships or links between geometric entities and functional requirements (Chapter 4). Algorithms were developed for the determination of boundedness and the measurement of sufficiency (Chapter 5). A detailed constraint network was used to represent tolerance relations for a part under design and provide for the composition of tolerance specifications (Chapter 6). Query methods for the constraint network were developed to support the measurement of tolerance sufficiency and tolerance synthesis through optimization (Chapter 7). To demonstrate the improved facilities for tolerance specification which result from the integration and interaction within the framework, CASCADE-T, a prototype computing environment, has been implemented (Chapter 8).

9.1 Computer Methods for Tolerance Design

The contributions of this work can be seen in five different areas : primitives for tolerance representation, methods for design composition, an approach for determining tolerance sufficiency, techniques for querying and explaining detailed design representations, and computer tools that support tolerance specification.

A set of tolerance primitives has been developed to model assembly requirements for common geometric features. As well, the approach used for geometric features has been extended to allow other types of performance requirements to be represented with tolerance primitives. Each tolerance primitive has particular functional requirements embedded within it that are complete by definition. That is, each functional requirement implies certain geometric relationships and the presence of these geometric relationships provides sufficient evidence that the functional requirement is satisfied. In the framework for tolerance synthesis, tolerance primitives provide a means for tolerance representation and for linking functional requirements to part geometry.

A definition for the degrees of freedom in a tolerance specification has been proposed and used to develop algorithms for the determination of boundedness and measurement of sufficiency. An inspection procedure was first developed for the approach defined by Requicha [53]. A second procedure, which exploits the detailed representation and tolerance theory employed by the framework for tolerance synthesis, was then developed. Both inspection procedures incur significant running time for tolerance specifications that are sufficient. The procedure defined for the framework for tolerance synthesis is superior in that it provides a measure of validity and sufficiency while reducing the computation required to determine fixed positions.

Composition of tolerance primitives has been investigated and several methods for composition and consistency management have been developed. Cases of independent and coupled functional requirements were defined and approaches were described to support composition for each case. For independent functional requirements, the framework for tolerance synthesis provides for automatic and interactive evaluation of validity for tolerance specifications. Competing requirements can be intersected automatically to maintain validity during the design effort. When requirements are contradictory, the designer is supplied with information about all parameters and equations that contribute to the conflict. For coupled functional requirements, additional checking must be done for each tolerance primitive and further analysis may be required to provide appropriate conditional tolerance relationships. Still, the framework for tolerance synthesis provides the majority of the representation and computing procedures required for these cases. These methods support tolerance specification procedures in a manner similar to the procedures currently available for building complex geometric solids from combinations of primitive solids.

As tolerance primitives and shape descriptions are composed, the design representation becomes very detailed. Methods for managing the complexity of this detailed representation have also been developed. An approach for the explanation of constraint networks was proposed as a means of providing compact and comprehensible views of the design representation. A hierarchical structure, where control and domain knowledge are separate, was employed so that explanations could be tailored to particular domains. According to the perspective and detail required, different but consistent types of explanations can be generated. As well, these explanations are useful for generating local optimization formulations. Further, the representation employed for the framework for tolerance synthesis was shown to be effective for

9.1. COMPUTER METHODS FOR TOLERANCE DESIGN

the generation of optimization formulations which can be used for final synthesis of tolerance specifications.

Finally, the methods and algorithms developed in this work have been implemented in a computer environment for tolerance specification. A library of tolerance primitives has been implemented by using the IDEEA environment and its constraint definition facility. Following the Application Interface Specification (AIS), an interface between a solid modeler and the IDEEA environment has been implemented. Constraint propagation is used to check the validity of tolerance specifications each time additions or modifications are made to the design. Contradictions are signaled to the user and explanations of the constraint network may be generated to suggest design changes that will be consistent with the functional requirements described by the tolerance primitives. The computer implementation provides a concrete demonstration of this approach to tolerance specification and gives useful insight into the scale of computation that is necessary to support such a detailed design representation.

Using the results of this book, current CAD environments can be extended significantly to provide intelligent design tool for assemblies of mechanical and electronic components. Design errors and manufacturing costs can be reduced by automatically detecting tolerancing errors and determining cases where tolerances can be relaxed. In addition, the detailed tolerance specification provided by the framework for tolerance synthesis provides a consistent method for the sharing of engineering knowledge among designers and manufacturing engineers. From the detailed account, several different but equally accurate interpretations of the part can be presented.

The approach of this book can also be used for other engineering problems and to enhance existing computer-aided engineering systems. Computer-based tools are commonly developed to enforce validity of prespecified design features. While these systems are very useful for analysis, they do not adequately support synthesis. When using these systems a designer is always faced with looking for functional requirements that will coincide with valid design features. In effect, the designer is asked to find a design problem that has one of the specified solutions. When relationships between functional requirements and design features can be expressed with equations or sets, this analysis-based paradigm can be extended by using the techniques developed for the framework for tolerance synthesis. The detailed representation that is employed can be used to assist designers in determining which solutions are consistent with the design problem under consideration. Further manipulation of the constraint network can be then be performed to choose an optimal design while at the same time remaining true to the original functional requirements.

We have presented several new techniques for tolerance design that allow for integration among many engineering disciplines. While the discussion and examples were generally posed with regard to geometric tolerances, these techniques are applicable to all types of design margins and tolerances.

Tolerances and design margins are applied in all engineering decisions to accommodate the uncertainty that is inherent in engineering practice. The techniques we have presented here are applicable to all types of design margins and tolerances. As well, our approach addresses very well the new demands of concurrent engineering. The tolerances and margins expressed in individual designs serve as the communication medium for negotiation among the concurrent engineering team.

9.2 Open Areas for Research

The five areas of contribution in this book each provide many opportunities for future work. Extensions are discussed below for each of the areas. Several new areas of research such as primitives for sound datum systems are also suggested.

9.2.1 Tolerance Representation

The tolerance primitives, described in Chapter 4, record information about tolerance assertion parameters, datums, fitting parameters, and conditional tolerance relations. The coupled case identified for the composition of tolerance primitives suggests that further information should be retained for each tolerance primitive. Coupled cases could be detected more generally if supporting assumptions for each conditional tolerance requirement were included with each tolerance primitive.

This type of information would also be quite useful for determining the soundness of the datum system defined for a particular tolerance specification. In fact, a simpler variety of tolerance primitives which only address datum systems would be readily applicable to present CAD tools. Reasoning strictly about datum systems appears to be a research area where significant results could be realized very quickly.

9.2.2 Sufficiency

The approach for sufficiency, described in Chapter 5, includes the assumption that all features of a part are constrained by some functional requirement. In the limit this will always be true since the variation of each unconstrained feature is eventually determined by the manufacturing process used to realize the feature. In a manufacturing environment where quality control is practiced, the variations will always be bounded quite well. In early stages of design, however, it may be useful to provide default variational classes for each nominal solid and then allow the functional requirements of a particular design to displace the constraints posed by the default variational classes.

9.2.3 Validity and Composition

The examples shown in Chapter 6 suggest that it would be very difficult to define general case algorithms to handle cases where functional requirements are coupled. For well-contained sets of design primitives, however, it should be possible to define a

9.2. OPEN AREAS FOR RESEARCH

priori all combinations of primitives which are consistent. As well, it may be possible to define consistent combinations of assumptions and datum systems.

The approach used to check the validity of independent functional requirements should also be useful for the selection of unspecified tolerances and for make or buy decisions. In current practice, unspecified tolerances are often determined according to defaults associated with the drawing or the fabrication process. Valid default tolerances could be selected automatically and checked via constraint propagation by considering the tolerances of available manufacturing processes and possible vendors. When in-house manufacturing equipment supplies tolerances that are consistent with other functional requirements, planning for in-house manufacturing should commence. Otherwise a procurement task could be initiated.

9.2.4 Explanation

While the rule-based approach of Chapter 7 can significantly reduce the detail and increase the focus of explanations, engineers often approach multi-variable problems by employing sensitivity analysis and graphical trade studies. It appears that the equations generated via explanations could quite readily be used to generate both sensitivity analyses and multi-variable trade studies.

As well, defining an appropriate set of rules and facts for explanation segregation and aggregation can be quite onerous. It appears that a great deal of this task could be automated or at least replaced by some sort of template for a set of rules and facts. A significant portion of the rules and facts might also be generated automatically according to an ordered set of preferences similar to those defined for conflict resolution [84].

Several refinements could easily be made to reduce the complexity of the explanation computations. The marking functions used to generate verbose explanations could be employed to add state information to the constraint network. In a more robust rule-based implementation, it would be possible to evaluate the constraint network and the explanation state directly via rule evaluation. The use of backward chaining rules would dramatically reduce the complexity of determining terminal nodes and attributes such as degrees of freedom.

A parser-based approach to the explanation task also appears promising. Instead of using rules and facts, it should be possible to develop sets of grammars for different segregation and aggregation policies.

Further refinements are also possible for the optimization formulations discussed in Chapter 7. For tolerance specification problems, it would be very useful to develop a means of generating tightly bounded formulations when intervals are part of the constraining equations. Further, an automated means could be developed to use minimization techniques to calculate unbound values and then introduce the results back into constraint network.

9.2.5 Extending CAD Tools

While the CASCADE-T system represents a significant step in the efficient use of a complex and consistent design representation for tolerance specification, more significant advances can be made by extending general CAD tools. Several improvements will be necessary to accomplish this type of extension. Dynamic data exchange must be provided between the CAD environment and accompanying computing processes for tolerance specification. Note that, in this type of application, the computing resources required for tolerance specification may be much larger than those required for the CAD environment.

As well, the specification procedure for tolerance primitives can be greatly improved by automating repetitive steps. The examples shown in Chapter 8 required a significant amount of user interaction to specify links between functional requirements and nominal solid geometry. For particular combinations of geometric features and tolerance primitives, it is expected that parts of this task could be assisted or handled by mechanized knowledge sources operating on functional requirements and nominal solid geometry.

Finally, since the constraint-based computing used in our approach is applicable to many different engineering tasks, it appears that a computing server approach might be useful for developing constraint-based computing machines. Note that computing speed is only one important aspect of this server. Improvements in communications bandwidth and protocols will also be necessary to accommodate the contradictions and resolutions of the constraint-based approach to computing.

Appendix A

Tolerance Primitive Definitions

(defun assembly—cyl—stud—orient—tol—constraints
  (aframe &aux net1 net2 flag)
  "applies orientation tolerance constraints to frame aframe to
enforce assembly requirement for cylindrical stud fitting into
cylindrical hole (Property 7). Contraints include a test of the
necessary condition of instance diameter s1 being smaller than
nominal diameter {plus} tolerance (sN1 + 2a1). If the necessary
condition is not met, the assembly requirement conditional
tolerance is not used for calculations - its TMS state is forced
:OUT"

  (declare (ignore flag))
  (with—binding—contour
      ((frame aframe)
       (no—value (constant nil))))

   (constrain—binding—contour
    ((frame a1 c3 l1 s1 sn1))

    ;; setting each possible route to calculating c3
    (multiple—value—setq (flag net1)
       (constrain tolprim—sol.m(sn1 l1 a1 s1) = c3))

    (multiple—value—setq (flag net2)
       (constrain c3 = no—value))

    ;; initial state set to use conditional tol.
    (setf (net—in? net1) t (net—in? net2) nil)

```
(with-binding-contour
    ((neta (constant net1)) (netb (constant net2)))

    ;; chose calculation route according to necessary condition
    (constrain sn1 + 2 * abs(a1) = net?(neta netb s1)
            :network "Nec. Cond for tolprim-so1.m")))))
```

```
(defun bulk-cyl-hole-orient-tol-constraints (aframe)
  "applies orientation tolerance constraints to frame aframe to
enforce material bulk requirement for cylindrical hole holding a
cylindrical stud (Property 7).  Contraints include a test of the
necessary condition of instance diameter s1 being smaller than
nominal diameter {plus} tolerance (sN1 + 2a1). If the necessary
condition is not met, the material bulk requirement conditional
tolerance is not used for calculations - its TMS state is forced
:OUT"
  (assembly-cyl-stud-orient-tol-constraints aframe))
```

```
(defun assembly-cyl-hole-orient-tol-constraints
    (aframe &aux net1 net2 net3 flag)
  "applies orientation tolerance constraints to frame aframe to
enforce assembly requirement for cylindrical hole holding a
cylindrical stud (Property 11).  Constraints include a test of
the necessary condition of instance diameter s1 being greater
than nominal diameter {plus} tolerance (sN1 + 2a1). c3 is
calculated according to three cases :

case 1 : s1 < l1

case 2 : l1 < s1 and sqrt(s1 * s1 - l1 * l1) < sn1 - 2 * abs(a1)

case 3 : otherwise

If the necessary condition is not met, the assembly requirement
conditional tolerance is not used for calculations - its TMS
state is forced :OUT"

  (declare (ignore flag))
```

```
(with-binding-contour
 ((frame aframe)
  (no-value (constant nil))
  (upto-halfpi
   (embedded-value (make-i :low 0.0 :high (/ Pi 2)))))
 (constrain-binding-contour
  ((frame a1 c3 l1 s1 sn1 cond0 cond1 cond2 cond3 cond4))
  (constrain cond0 = s1 <? (sn1 - 2 * abs(a1)))
  (constrain cond1 = s1 >? (sn1 - 2 * abs(a1)))
  (constrain cond2 = s1 <? l1)
  (constrain cond3 = s1 >? l1)
  (constrain cond4 =
             (sqrt(s1 * s1 - l1 * l1)) <? (sn1 - 2 * abs(a1)))

  (multiple-value-setq (flag net1)
       (constrain c3 = soft=(no-value)))
  (multiple-value-setq (flag net2)
       (constrain c3 = soft=(upto-halfpi)))
  (multiple-value-setq (flag net3)
       (constrain tolprim-so2.m( sn1 l1 a1 s1 ) = c3))

  (setf (net-in? net1) nil
        (net-in? net2) nil
        (net-in? net3) nil)

  (with-binding-contour
   ((neta (constant net1))
    (netb (constant net2))
    (netc (constant net3)))

   (constrain neta = true-when( cond0 ))
   (constrain netb
              = true-when(
                          select-and(
                                     cond1
                                     select-or
                                     ( cond2
                                       select-and (
                                                   cond3
                                                   cond4)))))
   (constrain netc = true-when(
                          select-and( cond1 cond2)))
```

))))

(defun bulk—cyl—stud—orient—tol—constraints (aframe)
  "applies orientation tolerance constraints to frame aframe to
enforce material bulk requirement for cylindrical stud fitting
into cylindrical hole (Property 11). Constraints include a test
of the necessary condition of instance diameter s1 being greater
than nominal diameter {plus} tolerance (sN1 + 2a1). c3 is
calculated according to three cases :

case 1 : s1 < l1

case 2 : l1 < s1 and sqrt(s1 * s1 - l1 * l1) < sn1 - 2 * abs(a1)

case 3 : otherwise

If the necessary condition is not met, the material bulk
requirement conditional tolerance is not used for calculations -
its TMS state is forced :OUT"
  (assembly—cyl—hole—orient—tol—constraints aframe))

(defun assembly—cyl—stud—position—tol—constraints
    (aframe &aux net1 flag)
  "applies position tolerance constraints to frame aframe to
enforce assembly requirement for cylindrical stud fitting into
cylindrical hole (Property 12). Contraints include tests of
the necessary conditions of instance diameter s1 and length l1
greater than 0 and c3 between 0 and pi/2. If the necessary
conditions are not met, the assembly requirement conditional
tolerance is not used for calculations - its TMS state is forced
:OUT"
  (declare (ignore flag))
  (with—binding—contour
      ((frame aframe)
       (no—value (constant nil))

```
(half-pi (constant (/ pi 2))))

(constrain-binding-contour
  ((frame a1 c1 c2 c3 c4 l1 l2 s1 sn1 cn1 cn2
          cond0 cond1 cond2 cond3))
  (constrain cond0 = c3 <? half-pi)
  (constrain cond1 = c3 >? (constant 0))
  (constrain cond2 = s1 >? (constant 0))
  (constrain cond3 = l1 >? (constant 0))

(multiple-value-setq (flag net1)
    (constrain tolprim-sp1.m
               (a1 c1 c2 c3 c4 l1 l2 s1 cn1 cn2 ) = sn1))

(setf (net-in? net1) nil)

(with-binding-contour
  ((neta (constant net1))
   (constrain neta =
                  true-when(select-and
                           (select-and
                           (cond0 cond1)
                           select-and(cond2 cond3))))))))))
```

```
(defun bulk-cyl-hole-position-tol-constraints (aframe)
  "applies position tolerance constraints to frame aframe to
enforce material bulk requirement for cylindrical hole holding a
cylindrical stud, (Property 12). Contraints include tests of the
necessary conditions of instance diameter s1 and length l1 greater
than 0 and c3 between 0 and pi/2. If the necessary conditions are
not met, the assembly requirement conditional tolerance is not
used for calculations - its TMS state is forced :OUT"
  (assembly-cyl-stud-position-tol-constraints aframe))


(defun assembly-cyl-hole-position-tol-constraints
    (aframe &aux net1 net2 flag)
  "applies position tolerance constraints to frame aframe to
enforce assembly requirement for cylindrical hole holding a
```

cylindrical stud, (Property 13). Contraints include tests of
the necessary conditions of instance diameter s1 and length l1
greater than 0 and c3 between 0 and pi/2. Requirements are
assigned according to two cases :

case 1 : l1 >= s1
case 2 : li < s1 and c3 < asin (l1/s1)
case 3 : otherwise.

If the necessary conditions are not met, the
assembly requirement conditional tolerance is not used for
calculations - its TMS state is forced :OUT"
```
  (declare (ignore flag))
  (with—binding—contour
   ((frame aframe)
    (half—pi (constant (/ pi 2)))
    )
   (constrain—binding—contour
    ((frame a1 c1 c2 c3 c4 l1 l2 s1 sn1 cn1 cn2
            cond0 cond1 cond2 cond3 cond4 cond5 cond6 cond7))

    (constrain cond0 = c3 <? half—pi)
    (constrain cond1 = c3 >? (constant 0))
    (constrain cond2 = s1 >? (constant 0))
    (constrain cond3 = l1 >? (constant 0))

    (constrain cond4 = l1 >? s1)
    (constrain cond5 = l1 <? s1)
    (constrain cond6 = c3 <? asin( l1 / s1 ))
    (constrain cond7 = c3 >? asin( l1 / s1 ))

    (multiple—value—setq (flag net1)
         (constrain tolprim—sp2.m
                   (a1 c1 c2 c3 c4 l1 l2 s1 cn1 cn2 ) = sn1))

    (multiple—value—setq (flag net2)
         (constrain  tolprim—sp3.m
                   (a1 c1 c2 c3 c4 l1 l2 s1 cn1 cn2 ) = sn1))

    (setf (net—in? net1) nil (net—in? net2) nil)

    (with—binding—contour
```

```
        ((neta (constant net1))
         (netb (constant net2)))
        (constrain neta = true—when
                    (select—and
                     (select—and
                      (select—and(cond0 cond1)
                                  select—and(cond2 cond3))
                       select—or(cond4
                                  select—and(cond5 cond6)))))
        (constrain netb = true—when
                    (select—and
                     (select—and
                      (select—and(cond0 cond1)
                                  select—and(cond2 cond3))
                        cond7)))))))
```

```
(defun bulk—cyl—stud—position—tol—constraints (aframe)
    "applies position tolerance constraints to frame aframe to
enforce assembly requirement for cylindrical stud fitting into
cylindrical hole, (Property 13). Constraints include tests of
the necessary conditions of instance diameter s1 and length l1
greater than 0 and c3 between 0 and pi/2. Requirements are
assigned according to two cases :

case 1 : l1 >= s1
case 2 : l1 < s1 and c3 < asin (l1/s1)
case 3 : otherwise.

If the necessary conditions are not met, the assembly requirement
conditional tolerance is not used for calculations - its TMS
state is forced :OUT"
    (assembly—cyl—hole—position—tol—constraints aframe))
```

;;;=========== Slab/Slot

```
(defun assembly—prism—slab—orient—tol—constraints (aframe &aux net1 net2 flag)
    "applies orientation tolerance constraints to frame aframe to
enforce assembly requirement for prismatic slab fitting into
```

prismatic slot (Property 15). Constraints include a test of the
necessary condition of instance thickness s1 being smaller than
nominal thickness {plus} tolerance (sN1 + 2a1). If the necessary
condition is not met, the assembly requirement conditional
tolerance is not used for calculations - its TMS state is forced
:OUT"

```
(declare (ignore flag))
(with-binding-contour
    ((frame aframe)
     (no-value (constant nil)))

  (constrain-binding-contour
   ((frame a1 c3 l1 s1 sn1))

    ;; setting each possible route to calculating c3
    (multiple-value-setq (flag net1)
      (constrain tolprim-so1.m( sn1 l1 a1 s1 ) = c3))

   (multiple-value-setq (flag net2)
     (constrain c3 = no-value))

   ;; initial state set to use conditional tol.
   (setf (net-in? net1) t (net-in? net2) nil)

   (with-binding-contour
       ((neta (constant net1)) (netb (constant net2)))

     ;; chose calculation route according to necessary condition
     (constrain sn1 + 2 * abs(a1) = net?(neta netb s1)
             :network "Nec. Cond for tolprim-so1.m")))))
```


(defun bulk-prism-slot-orient-tol-constraints (aframe)
  "applies orientation tolerance constraints to frame aframe to
enforce material bulk requirement for prismatic slot holding a
prismatic slab (Property 15). Constraints include a test of the
necessary condition of instance thickness s1 being smaller than
nominal thickness {plus} tolerance (sN1 + 2a1). If the necessary
condition is not met, the material bulk requirement conditional
tolerance is not used for calculations - its TMS state is forced

:OUT"
  (assembly—prism—slab—orient—tol—constraints aframe))

(defun assembly—prism—slot—orient—tol—constraints
  (aframe &aux net1 net2 net3 flag)
  "applies orientation tolerance constraints to frame aframe to
enforce assembly requirement for prismatic slot holding a
prismatic slab (Property 17). Constraints include a test of the
necessary condition of instance thickness s1 being greater than
nominal thickness {plus} tolerance (sN1 + 2a1). If the necessary
condition is not met, the assembly requirement conditional
tolerance is not used for calculations - its TMS state is forced
:OUT"

  (declare (ignore flag))
  (with—binding—contour
      ((frame aframe)
       (no—value (constant nil)))

    (constrain—binding—contour
     ((frame a1 c3 l1 s1 sn1))

     ;; setting each possible route to calculating c3
     (multiple—value—setq (flag net1)
       (constrain tolprim—so2.m(sn1 l1 a1 s1) = c3))

     (multiple—value—setq (flag net2)
       (constrain c3 = no—value))

     ;; initial state set to use conditional tol.
     (setf (net—in? net1) t (net—in? net2) nil)

     (with—binding—contour
         ((neta (constant net1)) (netb (constant net2)))

       ;; chose calculation route according to necessary condition
       (constrain sn1 + 2 * abs(a1) = net?(netb neta s1)
                  :network "Nec. Cond for tolprim-so2.m")))))

APPENDIX A. TOLERANCE PRIMITIVE DEFINITIONS

(defun bulk—prism—slab—orient—tol—constraints (aframe)
   "applies orientation tolerance constraints to frame aframe to enforce material bulk requirement for prismatic slab fitting into prismatic slot (Property 17). Constraints include a test of the necessary condition of instance thickness s1 being greater than nominal thickness {plus} tolerance (sN1 + 2a1). If the necessary condition is not met, the material bulk requirement conditional tolerance is not used for calculations - its TMS state is forced :OUT"

   (assembly—prism—slot—orient—tol—constraints aframe))

Appendix B
Tolerance Sufficiency Examples

The two-dimensional tolerance analysis of Chapter 5 is shown here in greater detail to illustrate the tolerance sufficiency definitions. For the part shown in Figure B.1, points and lines are the features used to describe the object. There are no degrees of freedom among the parameters for the features of the part and there is no redundancy among the constraints.

1. The types of geometric constraints used in this tolerance specification [15] are reviewed in Table B.1.

2. The actual equations for the geometric constraints of this example are shown in Tables B.2–B.4.

3. Figure B.2 shows the labeling that was used to generate the equations.

    Points are labeled as P1 – P8. Each point has x and y coordinates. These coordinates are referenced by number in the equations.

    Lines are labeled as 9 – 16. Each line has a length l which is referenced by number in the equations.

    Geometric constraints are labeled as C17 – C33.

4. The coincident point constraints are implicitly addressed in Equations B.9-B.16.

5. All other constraints are defined in Equations B.17-B.26.

6. Tolerances, shown as t, in the equations are assigned for each line length and for each geometric constraint. The dimensions and tolerances are addressed in Equations B.27-B.33.

7. Figure B.3 shows a graph for the equations and variables of this example. The nodes of the graph are variables and equations. By counting the arcs connected to an equation, the number of included variables is determined. By counting the arcs connected to a variable, the number of equations involved is determined.

Table B.1: Review of Geometric Constraints

Implicit Form : The general equation of a line in implicit form is

$$ax + by + c = 0. \tag{B.1}$$

For a line expressed in terms of two points, (x_1, y_1) and (x_2, y_2), the implicit form can be written as

$$(x_2 - x_1)(y - y_1) = (y_2 - y_1)(x - x_1). \tag{B.2}$$

This form can be rearranged to yield

$$(y_2 - y_1)x + (x_1 - x_2)y + x_2 y_1 - x_1 y_2. \tag{B.3}$$

Parallel Lines : For two lines expressed in the implicit form, when $a_1 b_2 = a_2 b_1$ the lines are parallel. For two lines, between (x_1, y_1) and (x_2, y_2) and between (x_3, y_3) and (x_4, y_4), this can be expressed with respect to the endpoints of the lines as

$$\frac{(y_2 - y_1)}{(x_1 - x_2)} - \frac{(y_4 - y_3)}{(x_3 - x_4)}. \tag{B.4}$$

Perpendicular Lines : For two lines expressed in the implicit form, when $a_1 a_2 + b_1 b_2 = 0$ the lines are perpendicular. For two lines, between (x_1, y_1) and (x_2, y_2) and between (x_3, y_3) and (x_4, y_4), this can be expressed with respect to the endpoints of the lines as

$$\frac{(y_2 - y_1)}{(x_3 - x_4)} + \frac{(x_1 - x_2)}{(y_4 - y_3)} = 0. \tag{B.5}$$

Fixed Angle : For a line, expressed in terms of two points, (x_1, y_1) and (x_2, y_2), when the line is oriented relative to a grid, or coordinate system, at a fixed angle θ,

$$\tan^{-1}\left[\frac{(y_2 - y_1)}{(x_2 - x_1)}\right] = \theta. \tag{B.6}$$

Fixed Point : When a point (x_1, y_1) is located at a fixed point (x_f, y_f) relative to a grid, or coordinate system, the fixed point refers to the true position of each coordinate of the point. Any tolerance associated with the true position is allocated to each coordinate.

$$x_1 + \text{tolerance} = x_f \tag{B.7}$$

$$y_1 + \text{tolerance} = y_f \tag{B.8}$$

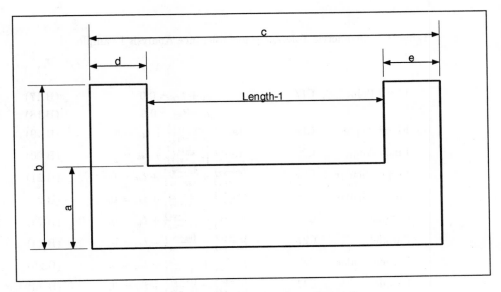

Figure B.1: Tolerance Analysis Example

Table B.2: Line Definitions for Tolerance Analysis Example

Line Definitions

$$l_9 = \sqrt{(x_1 - x_2)^2 + (y_1 - y_2)^2} + t_{34} \quad (B.9)$$

$$l_{10} = \sqrt{(x_2 - x_3)^2 + (y_2 - y_3)^2} + t_{35} \quad (B.10)$$

$$l_{11} = \sqrt{(x_3 - x_4)^2 + (y_3 - y_4)^2} + t_{36} \quad (B.11)$$

$$l_{12} = \sqrt{(x_4 - x_5)^2 + (y_4 - y_4)^2} + t_{37} \quad (B.12)$$

$$l_{13} = \sqrt{(x_5 - x_6)^2 + (y_5 - y_6)^2} + t_{38} \quad (B.13)$$

$$l_{14} = \sqrt{(x_5 - x_7)^2 + (y_6 - y_7)^2} + t_{39} \quad (B.14)$$

$$l_{15} = \sqrt{(x_7 - x_8)^2 + (y_7 - y_8)^2} + t_{40} \quad (B.15)$$

$$l_{16} = \sqrt{(x_8 - x_1)^2 + (y_8 - y_1)^2} + t_{41} \quad (B.16)$$

Table B.3: Geometric Tolerances for Tolerance Analysis Example

Fixed Point	C17	$x_1 + t_{42} = 0.0$	(B.17)
		$y_1 + t_{43} = 0.0$	(B.18)
Fixed Angle	C19	$\tan^{-1}\left[\frac{(y_8-y_1)}{(x_8-x_1)}\right] + t_{44} = 0.$	(B.19)
Fixed Angle	C20	$\tan^{-1}\left[\frac{(y_2-y_1)}{(x_2-x_1)}\right] + t_{45} = \frac{\pi}{2}$	(B.20)
Perpendicular	C21	$\frac{(y_1-y_2)}{(x_2-x_3)} + \frac{(x_2-x_1)}{(y_3-y_2)} + t_{46} = 0.$	(B.21)
Perpendicular	C25	$\frac{(y_4-y_3)}{(x_1-x_8)} + \frac{(x_3-x_4)}{(y_8-y_1)} + t_{47} = 0.$	(B.22)
Parallel	C26	$\frac{(y_2-y_1)}{(x_1-x_2)} - \frac{(y_7-y_8)}{(x_8-x_7)} + t_{48} = 0.$	(B.23)
Parallel	C27	$\frac{(y_5-y_4)}{(x_4-x_5)} - \frac{(y_8-y_1)}{(x_1-x_8)} + t_{49} = 0.$	(B.24)
Perpendicular	C28	$\frac{(y_5-y_6)}{(x_1-x_8)} + \frac{(x_6-x_5)}{(y_8-y_1)} + t_{50} = 0.$	(B.25)
Parallel	C32	$\frac{(y_7-y_6)}{(x_6-x_7)} - \frac{(y_8-y_1)}{(x_1-x_8)} + t_{51} = 0.$	(B.26)

Table B.4: Dimensions and Default Tolerances for Tolerance Analysis Example

Dimensions Specified	$l_9 = 3.0$	(B.27)
	$l_9 - l_{11} = 1.5$	(B.28)
	$l_{10} = 1.0$	(B.29)
	$l_{14} = 1.0$	(B.30)
	$l_{15} = 3.0$	(B.31)
	$l_{16} = 6.0$	(B.32)
Default Tolerances	$t_{34}, t_{35}, \ldots, t_{51} = 0.01$	(B.33)

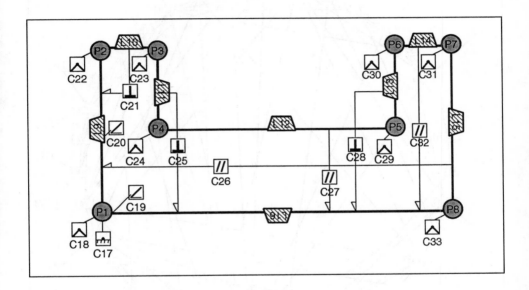

Figure B.2: Variable Labels for Tolerance Analysis Example

APPENDIX B. TOLERANCE SUFFICIENCY EXAMPLES

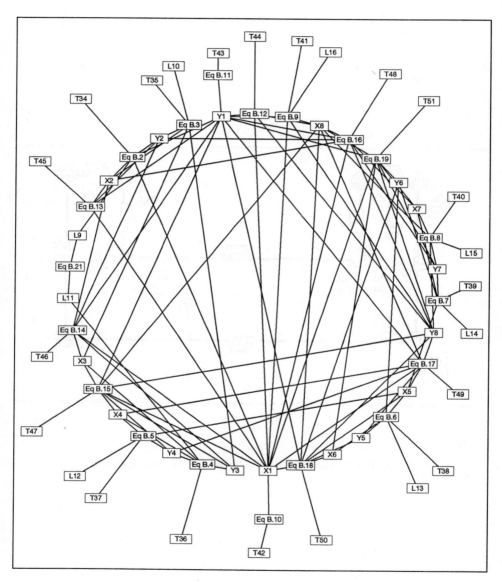

Figure B.3: Partial Graph for Tolerance Analysis Example

Appendix C

Constraint Explanation Examples

Tables C.1 – C.7 show the complete trace for a verbose explanation of how the orientation parameter c3 can be calculated for an instance of a cylindrical orientation tolerance primitive. Figure C.1 provides a graphical representation of the explanation.

Table C.1: Trace of Tolerance Primitive Constraint Explanation (1 of 7)

```
#<SLOT: C3 >   DETERMINED-BY
SETTING  #<SLOT: C3 >
#<SLOT: C3 >   COULD-BE-CALCULATED-WITH
<#:*3725:MULTIPLIER>   USING  *
M1  ALWAYS  #<SLOT: CELL-371 >  2.0
M2
#<SLOT: CELL-372 >   COULD-BE-CALCULATED-WITH
<#:ATAN3724:TAN>   USING  ATAN
TAN+1
#<SLOT: CELL-369 >   COULD-BE-CALCULATED-WITH
<#:/3723:MULTIPLIER>   USING  /
PRODUCT
#<SLOT: CELL-365 >   COULD-BE-CALCULATED-WITH
<#:/3708:MULTIPLIER>   USING  /
PRODUCT
#<SLOT: CELL-323 >   COULD-BE-CALCULATED-WITH
<#:|-3707|:ADDER>   USING  -
SUM
#<SLOT: CELL-320 >   COULD-BE-CALCULATED-WITH
<#:|+3706|:ADDER>   USING  +
X  #<SLOT: CELL-318 >   DETERMINED-BY
SETTING  #<SLOT: SN1 >
#<SLOT: SN1 >   COULD-BE-CALCULATED-WITH
<#:TOLPRIM-SO1.M7685:TOLPRIM-SO1.P>   USING  SN1
A1  #<SLOT: CELL-307 >   DETERMINED-BY
SETTING  #<SLOT: A1 >
#<SLOT: A1 >   COULD-BE-CALCULATED-WITH
<#:ABS3704:ABSOLUTE>   USING  ABS
PNUM  #<SLOT: CELL-313 >   DETERMINED-BY
#<SLOT: CELL-316 >   COULD-BE-CALCULATED-WITH
<#:*3705:MULTIPLIER>   USING  /
PRODUCT  #<SLOT: CELL-314 >   DETERMINED-BY
#<SLOT: CELL-319 >   WHICH-CYCLES
M1  ALWAYS  #<SLOT: CELL-315 >  2.0
#<SLOT: A1 >   COULD-BE-CALCULATED-WITH
<#:ABS3709:ABSOLUTE>   USING
ABS
```

Table C.2: Trace of Tolerance Primitive Constraint Explanation (2 of 7)

```
PNUM  #<SLOT: CELL-327 >   DETERMINED-BY
#<SLOT: CELL-330 >   COULD-BE-CALCULATED-WITH
<#:*3710:MULTIPLIER>   USING
/
PRODUCT  #<SLOT: CELL-328 >   DETERMINED-BY
#<SLOT: CELL-333 >   COULD-BE-CALCULATED-WITH
<#:|+3711|:ADDER>   USING
-
SUM  #<SLOT: CELL-331 >   DETERMINED-BY
#<SLOT: CELL-335 >   COULD-BE-CALCULATED-WITH
<#:|+3712|:ADDER>   USING
-
SUM  #<SLOT: CELL-334 >   DETERMINED-BY
#<SLOT: CELL-337 >   COULD-BE-CALCULATED-WITH
<#:/3713:MULTIPLIER>   USING
*
M1  #<SLOT: CELL-338 >   DETERMINED-BY
SETTING  #<SLOT: L1 >
#<SLOT: L1 >   COULD-BE-CALCULATED-WITH
<#:/3718:MULTIPLIER>   USING
/
PRODUCT
#<SLOT: CELL-351 >   COULD-BE-CALCULATED-WITH
<#:|-3717|:ADDER>   USING
-
SUM
#<SLOT: CELL-348 >   COULD-BE-CALCULATED-WITH
<#:|+3716|:ADDER>   USING
+
X  #<SLOT: CELL-346 >   COULD-BE-CALCULATED-WITH
<#:|+3716|:ADDER>   USING
-
SUM  #<SLOT: CELL-345 >   DETERMINED-BY
#<SLOT: CELL-348 >   WHICH-CYCLES
Y  #<SLOT: CELL-347 >   DETERMINED-BY
#<SLOT: CELL-342 >   WHICH-CYCLES
Y
```

Table C.3: Trace of Tolerance Primitive Constraint Explanation (3 of 7)

```
#<SLOT: CELL-347 >   COULD-BE-CALCULATED-WITH
<#:*3715:MULTIPLIER>   USING
*
M1   ALWAYS   #<SLOT: CELL-343 >   2.0
M2
#<SLOT: CELL-344 >   COULD-BE-CALCULATED-WITH
<#:ABS3714:ABSOLUTE>   USING
ABS
NUM   #<SLOT: CELL-340 >   COULD-BE-CALCULATED-WITH
<#:ABS3714:ABSOLUTE>   USING
ABS
PNUM   #<SLOT: CELL-341 >   DETERMINED-BY
#<SLOT: CELL-344 >   WHICH-CYCLES
X   #<SLOT: CELL-349 >   DETERMINED-BY
SETTING   #<SLOT: S1 >
#<SLOT: S1 >   COULD-BE-CALCULATED-WITH
<#:NET?7701:NET?>   USING
TESTPT
NET1   ALWAYS   #<SLOT: CELL-389 >   <CNET:#:CNET7692 is IN>
NET2   ALWAYS   #<SLOT: CELL-390 >   <CNET:#:CNET7693 is OUT>
VAL   #<SLOT: CELL-392 >   DETERMINED-BY
#<SLOT: CELL-386 >   COULD-BE-CALCULATED-WITH
<#:<?3615:<?>   USING
>?<
VAL1   #<SLOT: CELL-396 >   COULD-BE-CALCULATED-WITH
<#:<?3615:<?>   USING
>?<
VAL2   #<SLOT: CELL-397 >   DETERMINED-BY
#<SLOT: CELL-392 >   #<SLOT: CELL-386 >   WHICH-CYCLES
BOOL   #<SLOT: CELL-398 >   DETERMINED-BY
#<SLOT: CELL-395 >   WHICH-CYCLES
BOOL   #<SLOT: CELL-398 >   DETERMINED-BY
#<SLOT: CELL-395 >   COULD-BE-CALCULATED-WITH
<#:<NET-SWITCH3614:NET-SWITCH>   USING
SWITCH
NET1   ALWAYS   #<SLOT: CELL-393 >   <CNET:#:CNET7692 is IN>
NET2   ALWAYS   #<SLOT: CELL-394 >   <CNET:#:CNET7693 is OUT>
```

Table C.4: Trace of Tolerance Primitive Constraint Explanation (4 of 7)

```
#<SLOT: CELL−386 >  COULD−BE−CALCULATED−WITH
<#:|+7700|:ADDER>  USING
+
X  #<SLOT: CELL−387 >  COULD−BE−CALCULATED−WITH
<#:|+7700|:ADDER>  USING
−
SUM  #<SLOT: CELL−386 >  DETERMINED−BY
#<SLOT: CELL−392 >  #<SLOT: CELL−397 >  WHICH−CYCLES
Y  #<SLOT: CELL−388 >  DETERMINED−BY
#<SLOT: CELL−383 >  WHICH−CYCLES
Y
#<SLOT: CELL−388 >  COULD−BE−CALCULATED−WITH
<#:*7699:MULTIPLIER>  USING
*
M1  ALWAYS  #<SLOT: CELL−384 >   2.0
M2
#<SLOT: CELL−385 >  COULD−BE−CALCULATED−WITH
<#:ABS7698:ABSOLUTE>  USING
ABS
NUM  #<SLOT: CELL−381 >  COULD−BE−CALCULATED−WITH
<#:ABS7698:ABSOLUTE>  USING
ABS
PNUM  #<SLOT: CELL−382 >  DETERMINED−BY
#<SLOT: CELL−385 >  WHICH−CYCLES
M2  #<SLOT: CELL−353 >  DETERMINED−BY
#<SLOT: CELL−356 >  COULD−BE−CALCULATED−WITH
<#:*3719:MULTIPLIER>  USING
/
PRODUCT  #<SLOT: CELL−354 >  DETERMINED−BY
#<SLOT: CELL−359 >  COULD−BE−CALCULATED−WITH
<#:|+3720|:ADDER>  USING
−
SUM  #<SLOT: CELL−357 >  DETERMINED−BY
#<SLOT: CELL−361 >  COULD−BE−CALCULATED−WITH
<#:SQRT3721:SQR/RT>  USING
SQR
XX  #<SLOT: CELL−360 >  DETERMINED−BY
```

Table C.5: Trace of Tolerance Primitive Constraint Explanation (5 of 7)

```
#<SLOT: CELL-364 >  COULD-BE-CALCULATED-WITH
<#:|+3722|:ADDER>   USING
-
SUM  #<SLOT: CELL-362 >  DETERMINED-BY
#<SLOT: CELL-366 >  WHICH-CYCLES
X   ALWAYS  #<SLOT: CELL-363 >  1.0
X   ALWAYS  #<SLOT: CELL-358 >  1.0
M1  #<SLOT: CELL-355 >  DETERMINED-BY
#<SLOT: CELL-339 >  WHICH-CYCLES
M2  #<SLOT: CELL-339 >  COULD-BE-CALCULATED-WITH
<#:/3713:MULTIPLIER>   USING
/
PRODUCT  #<SLOT: CELL-337 >  DETERMINED-BY
#<SLOT: CELL-334 >  WHICH-CYCLES
M1  #<SLOT: CELL-338 >  DETERMINED-BY
#<SLOT: L1 >  #<SLOT: CELL-309 >  #<SLOT: CELL-324 >
#<SLOT: CELL-352 >  WHICH-CYCLES
Y   #<SLOT: CELL-336 >  COULD-BE-CALCULATED-WITH
<#:|+3712|:ADDER>   USING
-
SUM  #<SLOT: CELL-334 >  DETERMINED-BY
#<SLOT: CELL-337 >  WHICH-CYCLES
X   #<SLOT: CELL-335 >  DETERMINED-BY
#<SLOT: CELL-331 >  WHICH-CYCLES
X   #<SLOT: CELL-332 >  COULD-BE-CALCULATED-WITH
<#:|+3711|:ADDER>   USING
-
SUM  #<SLOT: CELL-331 >  DETERMINED-BY
#<SLOT: CELL-335 >  WHICH-CYCLES
Y   #<SLOT: CELL-333 >  DETERMINED-BY
#<SLOT: CELL-328 >  WHICH-CYCLES
M1  ALWAYS  #<SLOT: CELL-329 >  2.0
C3  #<SLOT: CELL-308 >  COULD-BE-CALCULATED-WITH
<#:TOLPRIM-SO1.M7685:TOLPRIM-SO1.P>   USING
C3
A1  #<SLOT: CELL-307 >  DETERMINED-BY
#<SLOT: A1 >  #<SLOT: CELL-312 >  #<SLOT: CELL-326 >
```

Table C.6: Trace of Tolerance Primitive Constraint Explanation (6 of 7)

```
#<SLOT: CELL-340 >  #<SLOT: CELL-381 >   WHICH-CYCLES
L1  #<SLOT: CELL-309 >   DETERMINED-BY
#<SLOT: L1 >  #<SLOT: CELL-324 >  #<SLOT: CELL-338 >
#<SLOT: CELL-352 >   WHICH-CYCLES
S1  #<SLOT: CELL-310 >   DETERMINED-BY
#<SLOT: S1 >  #<SLOT: CELL-321 >  #<SLOT: CELL-336 >
#<SLOT: CELL-349 >  #<SLOT: CELL-391 >  #<SLOT: CELL-396 >
WHICH-CYCLES
SN1  #<SLOT: CELL-311 >   DETERMINED-BY
#<SLOT: SN1 >  #<SLOT: CELL-318 >  #<SLOT: CELL-332 >
#<SLOT: CELL-346 >  #<SLOT: CELL-387 >   WHICH-CYCLES
L1  #<SLOT: CELL-309 >   COULD-BE-CALCULATED-WITH
<#:TOLPRIM-SO1.M7685:TOLPRIM-SO1.P>   USING
L1
A1  #<SLOT: CELL-307 >   DETERMINED-BY
#<SLOT: A1 >  #<SLOT: CELL-312 >  #<SLOT: CELL-326 >
#<SLOT: CELL-340 >  #<SLOT: CELL-381 >   WHICH-CYCLES
C3  #<SLOT: CELL-308 >   DETERMINED-BY
#<SLOT: CELL-5 >  #<SLOT: C3 >  #<SLOT: CELL-370 >
WHICH-CYCLES
S1  #<SLOT: CELL-310 >   DETERMINED-BY
#<SLOT: S1 >  #<SLOT: CELL-321 >  #<SLOT: CELL-336 >
#<SLOT: CELL-349 >  #<SLOT: CELL-391 >  #<SLOT: CELL-396 >
WHICH-CYCLES
SN1  #<SLOT: CELL-311 >   DETERMINED-BY
#<SLOT: SN1 >  #<SLOT: CELL-318 >  #<SLOT: CELL-332 >
#<SLOT: CELL-346 >  #<SLOT: CELL-387 >   WHICH-CYCLES
S1  #<SLOT: CELL-310 >   COULD-BE-CALCULATED-WITH
<#:TOLPRIM-SO1.M7685:TOLPRIM-SO1.P>   USING
S1
A1  #<SLOT: CELL-307 >   DETERMINED-BY
#<SLOT: A1 >  #<SLOT: CELL-312 >  #<SLOT: CELL-326 >
#<SLOT: CELL-340 >  #<SLOT: CELL-381 >   WHICH-CYCLES
C3  #<SLOT: CELL-308 >   DETERMINED-BY
#<SLOT: CELL-5 >  #<SLOT: C3 >  #<SLOT: CELL-370 >
WHICH-CYCLES
L1  #<SLOT: CELL-309 >   DETERMINED-BY
```

Table C.7: Trace of Tolerance Primitive Constraint Explanation (7 of 7)

```
#<SLOT: L1 >   #<SLOT: CELL-324 >   #<SLOT: CELL-338 >
#<SLOT: CELL-352 >   WHICH-CYCLES
SN1  #<SLOT: CELL-311 >   DETERMINED-BY
#<SLOT: SN1 >   #<SLOT: CELL-318 >   #<SLOT: CELL-332 >
#<SLOT: CELL-346 >   #<SLOT: CELL-387 >   WHICH-CYCLES
Y  #<SLOT: CELL-319 >   COULD-BE-CALCULATED-WITH
<#:|+3706|:ADDER>   USING
-
SUM  #<SLOT: CELL-317 >   DETERMINED-BY
#<SLOT: CELL-320 >   WHICH-CYCLES
X  #<SLOT: CELL-318 >   DETERMINED-BY
#<SLOT: SN1 >   #<SLOT: CELL-311 >   #<SLOT: CELL-332 >
#<SLOT: CELL-346 >   #<SLOT: CELL-387 >   WHICH-CYCLES
X  #<SLOT: CELL-321 >   COULD-BE-CALCULATED-WITH
<#:|-3707|:ADDER>   USING
-
SUM  #<SLOT: CELL-320 >   DETERMINED-BY
#<SLOT: CELL-317 >   WHICH-CYCLES
Y  #<SLOT: CELL-322 >   DETERMINED-BY
#<SLOT: CELL-323 >   WHICH-CYCLES
M1  #<SLOT: CELL-324 >   COULD-BE-CALCULATED-WITH
<#:/3708:MULTIPLIER>   USING
/
PRODUCT  #<SLOT: CELL-323 >   DETERMINED-BY
#<SLOT: CELL-322 >   WHICH-CYCLES
M2  #<SLOT: CELL-325 >   DETERMINED-BY
#<SLOT: CELL-365 >   WHICH-CYCLES
M1  #<SLOT: CELL-366 >   COULD-BE-CALCULATED-WITH
<#:/3723:MULTIPLIER>   USING
/
PRODUCT  #<SLOT: CELL-365 >   DETERMINED-BY
#<SLOT: CELL-325 >   WHICH-CYCLES
M2  #<SLOT: CELL-367 >   DETERMINED-BY
#<SLOT: CELL-369 >   WHICH-CYCLES
```

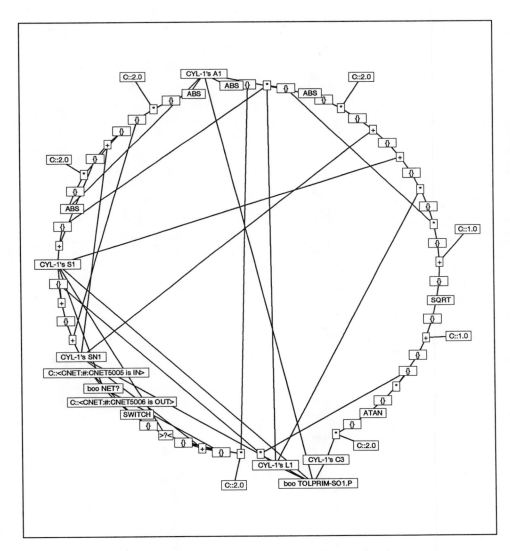

Figure C.1: Tolerance Primitive Constraint Graph

Bibliography

[1] American Society of Mechanical Engineering, New York, N.Y. *Dimensioning and Tolerancing ANSI Y14.5M.*

[2] A.A. Araya and S. Mittal. Compiling design plans from description of artifacts and problem solving heuristics. In *International Joint Conference on Artificial Intelligence*, pages 552–558, 1987.

[3] Lee Alton Barford. *A Graphical, Language-Based Editor for Generic Solid Models Represented by Constraints*. PhD thesis, Cornell University, 1987.

[4] Ø. Bjørke. *Computer-aided Tolerancing*. ASME Press, 1989.

[5] E. Charniak, C.K. Riesbeck, D.V. McDermott, and J.R. Meehan. *Artificial Intelligence Programming*. Lawrence Erlbaum Associates, second edition, 1987.

[6] Kenneth W. Chase, William H. Greenwood, and Bruce G. Loosli. Least cost tolerance allocation for mechanical assemblies with automated process selection. *Manufacturing Review*, 3(1), 1990.

[7] B.K. Choi, M.M. Barash, and D.C. Anderson. Automatic recognition of machined surfaces from a 3d solid model. *Computer-aided Design*, 16(2):81–86, 1984.

[8] CIRP. *The International Conference on CAD/CAM and AMT : CIRP Sessions on Tolerancing for Function in a CAD/CAM Environment*, 1989.

[9] H.G. Conway. *Engineering Tolerances   A Study of Tolerances, Limits and Fits for Engineering Purposes, with Full Tables of All Recognized and Published Tolerance Systems*. Pitman and Sons, 1966.

[10] L. De Floriani and B. Falcidieno. A hierarchical boundery model for solid object representation. *ACM Transactions on Graphics*, 7(1):42–60, January 1988.

[11] Z. Dong and A. Soom. Automatic optimal tolerance design for related dimension chains. *Manufacturing review.*, 3(4), 1990.

[12] Faryar Etesami. Tolerance verification through manufactured part modeling. *Journal of Manufacturing Systems*, 7(3):223–232, 1988.

[13] D. Fainguelernt, R. Weill, and P. Bourdet. Computer aided tolerancing and dimensioning in process planning. *Annals of CIRP*, 35(1):381–386, 1986.

[14] L.E. Farmer. Tolerancing for function in a cad/cam environment. In *Proceedings of the International Conference on CAD/CAM and AMT, CIRP Sessions on Tolerancing for Function in a CAD/CAM Environment*, pages (C-1-1)1-6, December 1989.

[15] I.D Faux and M.J. Pratt. *Computational Geometry for Design and Manufacture*. Ellis Horwood, 1979.

[16] A. Fleming. Geometric relationships between toleranced features. *Artificial Intelligence*, 37:403–412, 1988.

[17] E.T. Fortini. *Dimensioning for Interchangeable Manufacture*. Industrial Press, 1967.

[18] M.S. Fox, B. Allen, and G. Strohm. Job-shop scheduling : An investigation in constraint-directed reasoning. In *AAAI-82*, 1982.

[19] J.L. Gadzala. *Dimensional Control in Precision Manufacturing*. McGraw-Hill, 1959.

[20] D.C. Gossard, R.P. Zuffante, and H. Sakurai. Representing dimensions, tolerances, and features in MCAE systems. *IEEE Computer Graphics and Applications*, 8(2):51–59, March 1988.

[21] W.H. Greenwood and K.W. Chase. A new tolerance analysis method for designers and manufacturers. *Transactions of ASME, Journal of Engineering for Industry*, 109(2):112–116, May 1987.

[22] J.R. He. Toleranceing for manufacturing via cost minimization. *International Journal of Machine Tools and Manufacturing*, 31(4), 1991.

[23] A. E. Herman. An artificial intelligence based modeling environment for engineering problem solving. Master's thesis, University of Illinois at Urbana-Champaign, 1990.

[24] A.E. Herman and S. C-Y. Lu. A new modeling environment for developing computer-based intelligent associates with an application in machining operation planning. In *Transactions of the North American Manufacturing Research Institute of SME*, pages 260–265, May 1989.

BIBLIOGRAPHY

[25] R.C. Hillyard and I.C. Braid. Analysis of dimensions and tolerances in computer-aided mechanical design. *Computer-aided Design*, 10(3):161–166, May 1978.

[26] P. Hoffman. Automatic calculus of tolerances in discrete parts manufacturing. In A. Smith, editor, *Knowledge Engineering and Computer Modeling in CAD, Proceedings of CAD86*, pages 292–300, London, September 1986. Butterworths.

[27] R. Jayaraman and V. Srinivasan. Geometric tolerancing : I, Virtual boundery requirements. *IBM Journal of Research and Development*, 33(2):90–104, March 1989.

[28] S. Joshi and T.C. Chang. Graph-based heuristics for recognition of machined features from a 3d solid model. *Computer-aided Design*, 20(2):58–66, March 1988.

[29] S.H. Kim and K. Lee. An assembly modelling system for dynamic and kinematic analysis. *Computer aided design*, 21(1), 1989.

[30] S.V. Kulkarni and T.K. Garg. Allocation of tolerances to the components of an assembly for minimum cost. *Journal of the Institution of Engineers (India)*, 67(6):126–129, May 1987.

[31] W.-J. Lee and T.C. Woo. Optimum selection of discrete tolerances. *ASME Journal of Mechanisms, Transmissions, amd Automation in Design*, 111:243–251, 1989.

[32] W.-J. Lee and T.C. Woo. Tolerances: Their analysis and synthesis. *Journal of Engineering for Industry*, 112(2), 1990.

[33] Y.C. Lee and K.S. Fu. Machine understanding of csg : Extraction and unification of manufacturing features. *IEEE Computer Graphics and Applications*, pages 20–32, January 1987.

[34] A. Lehtihet and N.U. Gunasena. On the composite position tolerance for patterns of holes. *Annals of CIRP*, 40(1):495–498, 1991.

[35] E.A. Lehtihet and B.A. Dindelli. Tolocon: Microcomputer-based module for simulation of tolerances. *Manufacturing Review*, 2(3), 1989.

[36] R.W. Li. A part-feature recognition system for rotational parts. *International Journal of Production Research*, 26(9):1451–1475, 1988.

[37] S. C-Y. Lu. Multiple cooperative knowledge sources paradigm for concurrent product and process design: A step towards the knowledge processing technology, keynote paper. In *Proceedings of the 2nd Int. Conf. on Advanced Manufacturing Systems and Technology*, pages 130–143, Trento, Italy, July 1990.

[38] S. C-Y. Lu and R.G. Wilhelm. Applying constraint-based reasoning to geometric tolerancing: *Fifth Annual Conference on Artificial Intelligence in Engineering*, sponsored by the Computational Mechanics Institute, 1990..

[39] S. C-Y. Lu and R.G. Wilhelm. Automating tolerance synthesis : A framework and tools. *Journal of Manufacturing Systems*, 10(4):279–296, 1991.

[40] A.K. Mackworth. How to see a simple world : An exegesis of some computer programs for scene analysis. In E.W. Elcock and D. Michie, editors, *Machine Intelligence 8*, pages 510–537. Ellis Horwood Ltd., Wiley & Sons, 1977.

[41] S. Manivannan, A. Lehtihet, and P.J. Egbelu. A knowledge based system for the specification of manufacturing tolerances. *Journal of Manufacturing Systems*, 8(2):153–160, 1989.

[42] W. Michael and J.N. Siddall. The optimization problem with optimal tolerance assignment and full acceptance. *ASME Journal of Mechanical Design*, 103(4):842–849, 1981.

[43] W. Michael and J.N. Siddall. The optimal tolerance assignment with less than full acceptance. *ASME Journal of Mechanical Design*, 104(4):855–860, 1982.

[44] M. J. Muuss. Computer graphics techniques: Theory and practice. In D. F. Rogers and R. A. Earnshaw, editors, *Workstations, Networking, Distributed Graphics, and Parallel Processing*. Springer-Verlag, 1990.

[45] M. J. Muuss, P. Dykstra, K. Applin, G. Moss, P. Stay, and C. Kennedy. BRL CAD Package, release 3.0, a solid modeling system and ray-tracing benchmark. Technical report, Ballistic Research Laboratory, 1988.

[46] NASA Johnson Space Center. *CLIPS Reference Manual, Volume I, Basic Programming Guide*, 1991.

[47] Phillip F. Ostwald and Michael O. Blake. Estimating cost associated with dimensional tolerance. *Manufacturing Review*, 2(4), 1989.

[48] D.B. Parkinson. Tolerancing of component dimensions in cad. *Computer Aided Design*, 16(1):25–32, January 1984.

[49] R.J. Popplestone, A.P. Ambler, and I.M. Bellos. An interpreter for a language for describing assemblies. *Artificial Intelligence*, 14:79–107, 1980.

[50] M.J. Pratt. The CAM-I applications interface specification : Consolidated and restructured version, volume I, introduction and rationale. Technical report, Computer Aided Manufacturing – International, Inc., 1986.

[51] M.J. Pratt. The CAM-I applications interface specification : Consolidated and restructured version, volume II, – FORTRAN subroutine specification. Technical report, Computer Aided Manufacturing – International, Inc., 1986.

[52] P.S. Ranyak and R. Fridshal. Features for tolerancing a solid model. In *ASME Computers in Engineering Conference*, volume 1, pages 263–274, August 1988.

[53] A.G. Requicha. Toward a theory of geometric tolerancing. *The International Journal of Robotics Research*, 2(4):45–60, 1983.

[54] A.G. Requicha. Representation of tolerances in solid modeling : Issues and alternatives. In M.S. Pickett and J.W. Boyse, editors, *Solid Modeling by Computers*. Plenum, 1984.

[55] A.G. Requicha and S.C. Chan. Representation of geometric features, tolerances, and attributes in solid modelers based on constructive geometry. *IEEE Journal of Robotics and Automation*, RA-2(3):156–166, September 1986.

[56] U. Roy and C.R. Liu. Feature-based representational scheme of a solid modeler for providing dimensioning and tolerancing information. *Robotics and Computer-Integrated Manufacturing*, 4(3/4):335–345, 1988.

[57] H. Schmekel. Functional models and design solutions. *Annals of CIRP*, 38(1):129–132, 1989.

[58] J.J. Shah and D. Miller. A structure for supporting geometric tolerances for computer integrated manufacturing. In *Transactions of the North American Manufacturing Research Institution of SME*, pages 344–351, May 1989.

[59] R. Simmons. Commonsense arithmetic reasoning. In *AAAI-86*, pages 118–124, 1986.

[60] S. Srikanth and J.U. Turner. Toward a unified representation of mechanical assemblies. *Engineering With Computers*, 6:103–112, 1990.

[61] R. Srinivasan and C.R. Liu. Generative process planning using syntactic pattern recognition. *Computers in Mechanical Engineering*, 1984.

[62] V. Srinivasan and R. Jayaraman. Geometric tolerancing : II, Conditional tolerances. *IBM Journal of Research and Development*, 33(2):105–125, March 1989.

[63] S.M. Staley, M.R. Henderson, and D.C Anderson. Using syntactic pattern recognition to extract feature information from a solid geometric data base. *Computers in Mechanical Engineering*, pages 61–66, 1983.

[64] Guy L. Steele Jr. *The Definition and Implementation of a Computer Programming Language based on Constraints*. PhD thesis, M.I.T., 1980.

[65] K. Sugihara. Detection of structural inconsistency in systems of equations with degrees of freedom and its applications. *Discrete Applied Mathematics*, 10:297–312, 1985.

[66] Nam P. Suh. *The Principles of Design*. Oxford University Press, 1990.

[67] G.J. Sussman and G.L. Steel. Constraints — a language for expressing almost-hierarchical descriptions. *Artificial Intelligence*, 14:1–39, 1980.

[68] I.E. Sutherland. SKETCHPAD : A man-machine graphical communication system. Lincoln Laboratory Technical Report 296, Massachusetts Institute of Technology, January 1963.

[69] J.B. Thompson and S.C-Y. Lu. Representing and using design rationale in concurrent product and process design. In *ASME Winter Annual Meeting*, San Francisco, December 1989.

[70] V.V. Tipnis, editor. *Research Needs and Technological Opportunities in Mechanical Tolerancing*. American Society of Mechanical Engineers, 1990.

[71] P. Todd. An algorithm for determining consistency and manufacturability of dimensioned drawings. In A. Smith, editor, *Knowledge Engineering and Computer Modeling in CAD, Proceedings of CAD86*, pages 36–41, London, September 1986. Butterworths.

[72] Mark T. Traband, Sanjay Joshi, and Richard A. Wysk. Evaluation of straightness and flatness tolerances using the minimum zone. *Manufacturing Review*, 2(3):189–195, 1989.

[73] P. Treacy, J.B. Ochs, and T.M. Ozsoy. Automated tolerance analysis for mechanical assemblies modeled with geometric features and relational data structure. *Computer Aided Design*, 23(6):444, 1991.

[74] J.U. Turner and A.B. Gangoiti. Tolerance analysis approaches in commercial software. *Concurrent Engineering*, 1(2):11–23, March/April 1991.

[75] J.U. Turner and M.J. Wozny. A mathematical theory of tolerancing. In M.J. Wozny, H.W. McLaughlin, and J.L. Encarncao, editors, *Geometric Modeling for CAD Applications*. North Holland, 1988.

[76] G.N. Vanderplaats. *Numerical Optimization Techniques for Engineering Design : With Applications*. McGraw-Hill, 1984.

[77] R.J. Walsh. *Plain Bearing Design Handbook*. Butterworths, 1983.

[78] N. Wang and T.M. Ozsoy. Representation of assemblies for automatic tolerance chain generation. *Engineering with Computers*, 6(2), 1990.

BIBLIOGRAPHY

[79] A.C. Ward. *A Theory of Quantitative Inference for Artifact Sets, Applied to a Mechanical Design Compiler*. PhD thesis, Massachusetts Institute of Technology, 1989.

[80] R. Weill. Tolerancing for function. *Annals of CIRP*, 37(2):603–610, 1988.

[81] R. Weill. New methodologies toward the integration of tolerancing in cad/cam systems. In *Proceedings of the International Conference on CAD/CAM and AMT, CIRP Sessions on Tolerancing for Function in a CAD/CAM Environment*, pages (C–1–2)1–7, December 1989.

[82] R.G. Wilhelm. Generating useful explanations from constraint networks. Technical Report KBESRL-003-1989, Knowledge-Based Engineering Systems Research Laboratory, University of Illinois at Urbana-Champaign, 1989.

[83] R.G. Wilhelm. Geometric modeling with CLOS-AIS : an AIS implementation for lisp, phase I implementation. Technical Report KBESRL-CM-1991-002, Knowledge-Based Engineering Systems Research Laboratory, University of Illinois at Urbana-Champaign, 1991.

[84] R.G. Wilhelm and A.E. Herman. Computer aids for engineering design decisions : A simple contradiction handler for ideea applications. Technical Report KBESRL-CM-1991-003, Knowledge-Based Engineering Systems Research Laboratory, University of Illinois at Urbana-Champaign, 1991.

[85] R.G. Wilhelm and S. C-Y. Lu. Tolerance primitives for composition and synthesis. *Transactions of the North American Manufacturing Research Institute of SME*, pages 366–370, May 1990.

[86] R.G. Wilhelm and S. C-Y. Lu. Tolerances and function in concurrent product and process design. *Transactions of the North American Manufacturing Research Institute of SME*, pages 357–363, May 1991.

[87] Z. Wu, W.H. Elmaraghy, and H.A. Elmaraghy. Evaluation of cost–tolerance algorithms for design tolerance analysis and synthesis. *Manufacturing Review*, 1(3):168–179, 1988.

Index

admissible dimensioning, 14
aggregation, 19, 72
AIS, 86
Application Interface Specification, 86
assembly models, 15
assembly requirement, 28

bearing design
    example, 91
    Walsh, 42
bounded variational class, 50
bounding approximation, 40
BRL, 85

CAD extensions, 105, 108
CASCADE-T, 85, 105, 108
    design scenario, 85
    implementation, 86
CLIPS, 87
CLOS-AIS, 87, 95
coincident points, 119
composition, 55, 104
    coupled requirements, 60
        example, 67
    example, 100
    independent requirements, 58
        example, 61
conditional tolerance assertions, 34, 43
conditional tolerance zone
    example, 37
conflict resolution, 107
confounded parameters, 37
constraint definition
    example, 34, 37, 45, 61
constraint language, 17
    example, 87, 109

constraint networks, 14, 108
    contradiction, 63
        example, 87
    cycles, 63
    detecting cycles, 73
    explanation, 18, 72, 104
        example, 72
    graph
        example, 87
    IDEEA, 87
    queries, 72
    representation, 100
    traversal, 73
constraint propagation, 61, 63
    complexity, 67
constraint-based reasoning, 17
CONSTRAINTS language, 18
cost estimation, 16
cost minimization, 16
coupled conditional tolerances, 67
coupled functional requirement, 58
    example, 58

datum
    specification, 5
    surface, 4
datum systems, 106
default tolerances
    example, 122
default variational classes, 106
degrees of freedom, 50
design axioms, 58
design margin, 1
design synthesis, 19, 55
detailed design representation, 104

INDEX

dimension-driven geometry, 14

explanation, 18, 104
    aggregation example, 76
    complexity, 80, 107
    example, 72, 125
    graph, 133
    how, 72
    optimization formulations, 80, 107
    parser, 107
    why, 72
explanation hierarchy, 73
    control knowledge, 76
    domain knowledge, 76

feasible manufacturing material, 4
feasible manufacturing solid, 4
feature, 4
feature extraction, 5
feature recognition, 18
fitting parameters, 33, 43
fixed angle, 120
fixed point, 120
frame system, 87
framework for tolerance synthesis, 24
    composition, 55, 58
    constraint representation, 71
    explanation, 73
    functional requirements, 2
    integration, 85
    links to requirements, 27
    nominal solids, 2
    sufficiency, 50
    synthesis, 80
    tolerance primitives, 32
    tolerance representation, 28
    validity, 67, 69
functional gauge, 32
functional requirement, 2, 49
    coupled, 58
    independent, 56

geometric constraints, 119, 120

geometric tolerances, 1
    ANSI standard, 15
    design, 1
    example, 122
    fabrication, 1
geometric tolerancing
    current practice, 21

hierarchy for explanation, 73

IDEEA, 34, 86
implicit form, 120
independent functional requirement, 56
interval calculations, 87
ISIS-II, 17

material bulk, 28
Mechanical-Advantage, 49
Monte Carlo simulation, 16

nominal geometry, 2
nominal solids, 2, 27
    bounds, 51
    geometry, 2
    topology, 2

operating condition parameters, 43
optimization formulation, 80, 107
    example, 82

parallel lines, 120
part assembly
    example, 95
perpendicular lines, 120
picture interpretation, 18
Pro-Engineer, 49
process plan, 6

RAPT interpreter, 18
ROSCAT, 17
rule-based system, 87

segregation, 72
sensitivity analysis, 16, 49, 107
SKETCHPAD, 17

structural rigidity, 51
sufficiency, 47, 106, 119
    criteria, 50
    definition, 50
    graph, 119, 124
    inspection procedure, 52, 53
sufficiency measurement, 49, 104
syntactic pattern recognition, 18

theory of tolerancing, 15
tolerance analysis, 6, 13, 47
    example, 8, 48, 119
    sensitivity analysis, 14
tolerance assertion parameter, 33, 43
    bounds, 51
tolerance assertions, 27
tolerance chains, 14
tolerance errors
    consistency, 22
    sufficiency, 23
    validity, 22
tolerance feature identification, 4
tolerance primitive, 27, 32, 87, 103, 106
    assertion parameter, 33, 43
    bearing design, 42, 44
        constraint definition, 45
    bounding approximation, 40
    conditional tolerance assertions, 34, 43
    confounded parameters, 37
    consistency, 94
    consistent combinations, 107
    constraint network, 56
    cylindrical pairs, 33
    empirical requirements, 40
    example, 34, 67, 95, 109
    fitting parameters, 33, 43
    operating condition parameters, 43
    operating conditions, 40
    prismatic pairs, 33
tolerance specification, 4, 21
    complete, 5, 21
    composition, 26

invalid, 69
sufficient, 47, 50
valid, 56, 63
tolerance synthesis, 6, 15, 56
    computer-based, 23
    example, 7
    framework, 24
tolerance theory, 27
    nonparametric, 28
    parametric, 28
tolerances
    discrete, 16
tolerancing
    theory, 15
tolerancing for function, 16
trade studies, 107
truth-maintenance system, 87

unbounded variational class, 50

valid tolerance specification, 56
validity, 67, 69, 87, 104, 107
variational class, 50
    bounds, 51
    example, 100
VBR, 15
virtual boundary requirements, 15, 28
    necessary and sufficient conditions, 30
    parameterized features, 30
virtual half-spaces, 30
virtual surfaces, 28
    example, 28
    functional requirements, 28
VSA, 49

Walsh
    bearing design method, 42